Smart Metering Technologies

Edited by Inderpreet Kaur

Published in London, United Kingdom

IntechOpen

Supporting open minds since 2005

Smart Metering Technologies
http://dx.doi.org/10.5772/intechopen.91128
Edited by Inderpreet Kaur

Contributors
Doudou Nanitamo Luta, Akindele Olufemi Abegunde, Atanda Kamoru Raji, Inken Hoeck, Elmar Steurer, Biswajit Paul, Rajesh Palit, Khaled Mabrouk M Alkar, Inderpreet Kaur, Preeta Sharan, Manpreet Singh Singh Manna, Juriy Andreevich Tashkinov

First published in London, United Kingdom, 2021 by IntechOpen
IntechOpen is the global imprint of INTECHOPEN LIMITED, registered in England and Wales, registration number: 11086078, 5 Princes Gate Court, London, SW7 2QJ, United Kingdom
Printed in Croatia

British Library Cataloguing-in-Publication Data
A catalogue record for this book is available from the British Library

Additional hard and PDF copies can be obtained from orders@intechopen.com

Smart Metering Technologies
Edited by Inderpreet Kaur
p. cm.
Print ISBN 978-1-83969-355-7
Online ISBN 978-1-83969-356-4
eBook (PDF) ISBN 978-1-83969-357-1

For EU product safety concerns:
IN TECH d.o.o., Prolaz Marije Krucifikse Kozulić 3, 51000 Rijeka, Croatia, info@intechopen.com or visit our website at intechopen.com.

Meet the editor

Dr. Inderpreet Kaur has more than two decades of experience in research, education, and administration at higher-education institutes. She is currently a director for IGEN Edu Solutions, India. Her prime areas of interest are smart grids, OFC, and engineering education research. She has acquired fifty patents and has published numerous papers in international journals. She is also a reviewer of more than 100 international conferences. She is a board member for several national and international bodies, including the Institute of Electrical and Electronics Engineers (IEEE), the Malaysia Technical Scientist Association, and many more. Dr. Kaur is the recipient of awards such as the IEEE Appreciation Letter, India's Top 20 Women Leaders in Education 2020, Certificate of Appreciation 2019-IEEE Indonesia Section, and others.

Contents

Preface

This book covers the different methodologies and technologies used for smart metering technology (SMT). The first two letters in the term "smart" generally stand for specific and measurable. The term "art" is defined as something meaningful and beautiful that is produced with creativity. Thus, "smart" refers to parameters that are specific and measurable for producing something creative and meaningful to a user.

Metering is a process used to measure a quantity to calculate a result. The quantity could be an amount of gas, water, electricity, or voltage, or an exam score, and so on. The evaluation of the metering (measurement) process (EMP) in academics is a systematic calculation of a course's merit and its importance in a particular program. EMP usually follows a pre-defined set of criteria ruled by a set of standards. The main purpose of EMP is to enable reflection of the response of utility versus consumer or teacher versus learner and help in the identification of gaps. EMP should be a continuous monitoring process and must enable a two-way communication system through a feedback system or analyzers. For instance, in an academic system, the evaluation process is a continuous process that evaluates the progress of students through assignments, quizzes, mid-semester examinations (MSEs), and end-of-semester examinations.

The way or the methods people use to creatively organize tasks is known as technology. Technology must be user-friendly and have concern for human needs while keeping in mind the objectives of science. Technology must apply scientific knowledge and include practical knowledge as well as theoretical knowledge. Here, practical knowledge refers to facts, figures, analyses, and conclusions. Furthermore, technology involves organized ways of doing things. Technology involves interactions (both direct and indirect) between products (i.e., materials and processes), operators (i.e., people), and machines (i.e., systems). It implies that technology provides an interactive platform for people who either manufacture products or use products and byproducts throughout various stages of processes. For example, many people like to dine out. The ingredients of the meal (vegetables, fruits, coffee, tea, and other drinks, etc.) may come from plants, trees, gardens, and so on that have been cultivated in one region of the world but may require fertilizer or pesticides from another region. The harvested food must be transported and distributed to restaurants or hotels that process, cook, and serve the food to customers according to certain requirements and standards. Technology is necessary throughout the food supply chain to meet these requirements.

For academic users, SMT is a set of pre-defined methodologies used to evaluate the performance of learners (students). It may relate the EMP with the attendance of students. In some cases, students complete Exit Survey Forms (ESFs), providing feedback to improve the teaching-learning system. For instance, in one Indian institute of higher learning, students indicated in their ESFs that they would prefer to have two MSEs instead of three. After considering this concern, it was concluded that two MSEs would improve student performance. Thus, the evaluation system of that institute was revised from three MSEs to two MSEs.

Similarly, in terms of electrical energy, smart metering is the main initiative for fulfilling the objectives of smart grids. Smart grids solve critical problems related to energy, customer operations, peak management, and more. Undeniably, smart metering reduces commercial losses, monitors energy (in real-time or near real-time), detects energy theft, enhances grid reliability, and leads to better revenue as well as tariff management. It encourages consumers to change behavior and optimize energy consumption, even by generating electricity at their own premises and sending it back to the grid. Thus, smart metering architecture plays a vital role in achieving the objectives of a smart grid.

I hope that this book provides a good learning experience.

Inderpreet Kaur
Director,
IGEN Edu Solutions,
Chandigarh, India

Chapter 1

Low Cost, Robust and Multi-Functional Smart Meter

Akindele Olufemi Abegunde, Doudou Nanitamo Luta and Atanda Kamoru Raji

Abstract

Environmental awareness, current trends in the power market, the quest for energy efficiency, and the progressive transformation of electricity consumers to prosumers are the primary drives for the gradual shift from the old power grids into the smart grids. The deployment of renewable dispersed generation systems and energy storage units uncovered the need for smart metering to oversee and control those generation systems. This chapter presents the design and development of a robust, efficient, multi-functional, and low-cost smart meter. The proposed metering system has added features that enabled the utilities to recover the meter energy measurement data remotely. The system allows monitoring and transmission of energy consumed in real-time. It considers using a microcontroller board as the controlling unit to execute control and monitor activities. A liquid crystal display displays standard electrical measurements such as current, voltage, power, and energy consumption. The external communication device is required in the unit's actualization, in conjunction with the control unit based on the existing mobile technology. It stands as the intermediary between the nearby available utility station and consumers or end-users. In conclusion, liquid crystal display displays real-time based data for the end-user to visualize. The usage data billing is done within thirty seconds, stored, and trans-received the process for data collection, keeping, and billing generation.

Keywords: smart grid, smart meter, evolution of electricity meter, LCD, GSM

1. Introduction

In recent times, the deployment of renewable dispersed generation systems and energy storage units uncovered the need for smart metering to oversee and control the generating units. The first-generation of the smart meter was developed in 2005 to transmit data back to the energy supplier. During the process, transferring data every month was upgraded to sharing of data daily or hourly. The process has helped the customers to be able to consume and produce concurrently. This demonstrates smart meters' significance to electromechanical devices [1], which is only limited to electricity consumption measurement. References [2, 3] reported that in a year time (2020), an estimated one billion smart meters would be produced globally. The researcher further stated that the US would be closed to 65 million demands quota of smart meters by the said year: the expected highest demand by any country out a billion quantities. More so, dated as far back the year 1990, exploring gathered information collected from an energy metering device to bill through a central database came to limelight through a technology called Automatic Meter Reader over from then electromechanical meter.

The flowchart diagram displayed in **Figure 1** illustrates the process involved in smart meter evolution [2, 4]. Reference [5] stated that smart energy meter operates in two formats, such as the automatic meter reader (AMR) and the advanced meter infrastructure (AMI). According to Reference [6], AMR is an electronic meter that employs one-way communication data collection. It is a classy system that automatically calculates billing and relays the information about the energy supplier's consumption rate remotely. The system could involve various techniques to communicate, including general packet radio service (GPRS), supervisory control and data acquisition (SCADA), radiofrequency (RF), and global system for mobile (GSM). Given this, the researcher concluded that GSM is the most adaptive device with many users and the coverage zone for data transmission. This quality enhances the chances of using the system for metering purposes. Also, energy meters that use GSM prepare data for easy access to energy consumers and energy suppliers.

On the other hand, AMI is an electronic meter that communicates between the energy provider and customers by informing them about the specific interval data. AMI integrates two-way communication and an electronic meter designed to observe and regulate the grid system [7].

A smart metering system could be described as an energy system that measures energy consumption, data collection, data creation, and energy billing activities. References [6, 8–12] define smart meters as the device built and installed around a home or business to measure real-time consumption rate of electric, gas, and water used to envisage the improvement required for the accuracy, reliability, and efficiency enhancement of the outdated or/and overburden electrical, water and gas grids. Reference [13] categorically stated that a smart energy meter is an electrical device that tracks energy usage, and instantaneously communicates the energy supplier's outcome. Understandably, the process of transferring the energy captured, recorded, and stored at the electricity distributors through a wireless network takes ≤30 seconds to deliver. Reference [14] described the impact smart meter energy has on enhancing energy efficiency challenges through a concept called intelligent energy network. This concept comprises energy meter devices and intelligent communication technology (ICT). Intelligent energy networking was pointed

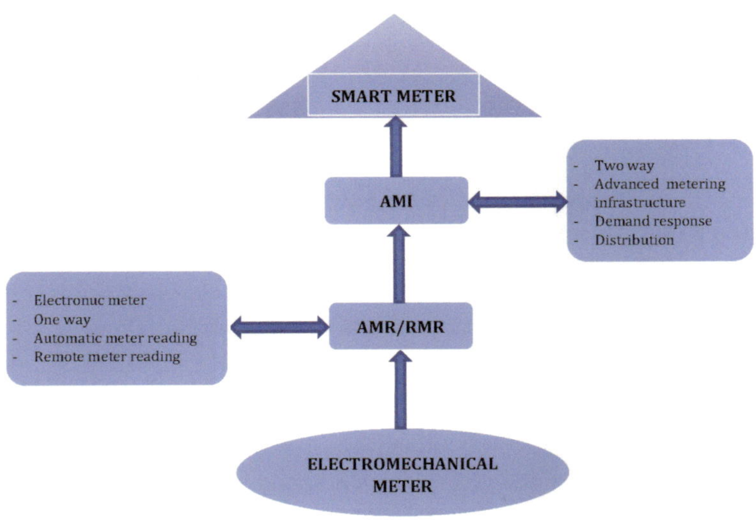

Figure 1.
Evolution of energy metering to the smart meter.

out as the ultimate energy device needed in achieving smart energy metering systems. This device can effectively monitor and control energy data exchange between the utility and the consumers. This process is performed in two-way directionally between meters to meters regarding the networking type imbibed. Reference [15] mentioned the significance of smart metering as an antidote to a more energy-efficient and metering system that gives accurate meter reading and billing system. However, smart metering has related working principles with the conventional meter in arrangement and calculation of physical quantities but differs from the computational aspect. Smart metering computes less energy consumption rate either in hourly or in seconds rather than in monthly. Reference [5] said that smart energy meter operates in two formats, such as AMR and AMI. AMR communicates and collects data for the utility company just in one direction. In the same section, AMI was described as an electronic meter that communicates between the energy provider and customers by informing them about the data collected at a certain interval. The further description illustrates that AMI integrates two-way communication and electronic meter to observe and regulate the grid system [7]. Additionally, a first-generation smart meter was developed in 2005 to transmit data back to the energy supplier. During the process, transmitting data on a monthly basis was upgraded to sharing data daily or hourly. The process has helped the customers to be able to consume and produce concurrently. This demonstrates smart meters' significance to electromechanical devices [16], which is only limited to electricity consumption measurement. Apart from that, the electromechanical device lacks consistency when it comes to energy measurement and encouragement for criminal activities. The demand for the supply of electrical energy brings about the existence of electronic meters with additional functions. However, electronic meters work on a principle of digital micro- technology (DMT). The application of this principle has no involvement in the moving disc, which results in wear and tear of the moving parts [17]. The electronic meter performs the automatic meter reading from consumers to both production and control executes by the utility. In that case, the smart energy meter combines the electronic device, intelligent communication technology, and control system in real time.

2. Architectural structure of smart meter

Although smart metering has related working principles with the conventional meter in the arrangement and calculation of physical quantities, they differ in the computational aspect. Smart metering computes less energy consumption rate either in hourly or in seconds rather than in monthly. **Figure 2** depicts a smart meter's general structure comprising two parts: hardware and software. The hardware part consists of three central units: acquisition, data processing, and data transmission units. These units represent the combination of components like a voltage sensor (VS), a current sensor (CS), an energy metering integrated circuit (EMIC), microcontroller unit (MCU), liquid crystal display (LCD), power supply/real-time clock (PS/RTC) and communication unit (CMU) [9, 18, 19].

2.1 Data acquisition unit

As one of the units considered in a smart meter's architectural development, data acquisition is referred to as a unit where analog data is obtained, processed, and converted into a required digital input for data processing. It is advised that careful execution of this process is necessary to generate a reliable result. This unit consists of the voltage sensor (VS), current sensor (CS), and level shifter circuits (LSC) [18].

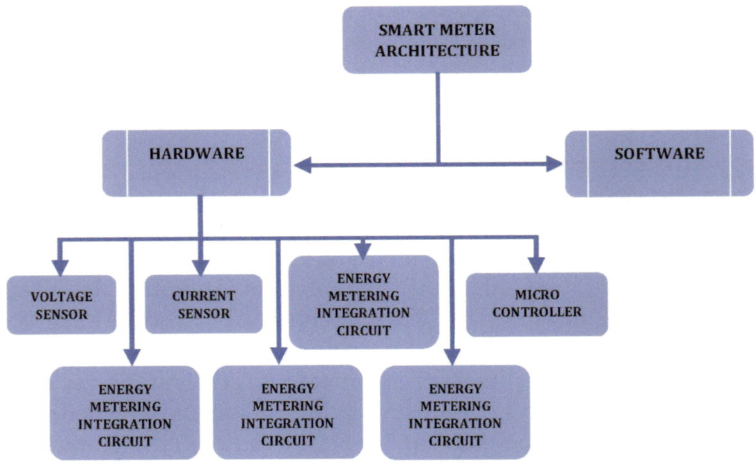

Figure 2.
Basic architecture of smart meter.

The VS and CS function as the facilitators of data acquisition before being transmitted to the energy metering integrated circuit (IC) for signal conditioning while simultaneously convert analog to digital developments. This type of controller is a "system on chip (SOC)." SOC constitutes analog front end (AFE) with a microcontroller unit (MCU). More so, AFE is a section of the smart energy device that is connected to the high voltage lines [18, 20]. This component regulates the high voltage and high current rates from the mains into smaller values ADC and MCU can easily absorb or process [21]. The MCU can be referred to as the device's brain because it dictates and controls all functions initiated within the smart energy meter.

2.2 Data transmission unit

The data transmission unit is responsible for transferring and receiving generated energy parameters to fully notify the billing and monitoring purposes to both the energy supplier and customers. Data is transmitted to a centralized server with customers' identities stored to determine the customers' unwillingness and criminal activities such as unpaid electricity usage, electricity theft, and electricity property vandalism [12].

3. Communication network system

Communication network systems for smart energy meters are the essential existing networks adapted into energy metering. It can be subdivided into cables and wireless networks, as shown in **Figure 3**. According to references [22, 23], a smart meter should be built to carry out functionalities like measuring, applying, and communicating energy parameters to stimulate efficiency and energy supply across households and industries. However, this efficiency is possible through a proper selection of communication networks and ports to manage energy data transmission and reception. Communication network systems must be cost-productive, give great transmittable extent, better security characteristics, data transmission, power quality, and the slightest conceivable number of repetitions.

Communication can be achieved using various communication procedures, including power line communication (PLC), ethernet, coaxial cable, RF, Wi-Fi, ZigBee, Bluetooth, GSM, and other available methods. The PLC carries data on

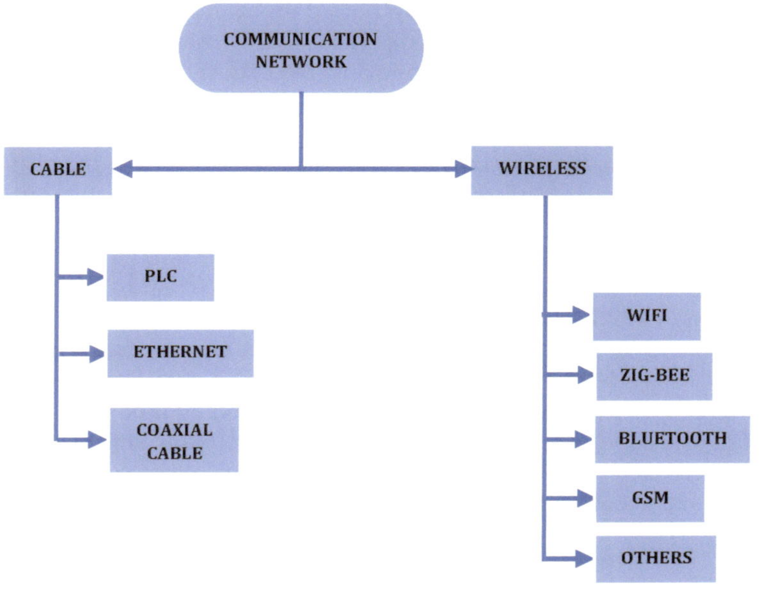

Figure 3.
Communication network systems for smart energy meters.

conductors employed simultaneously for AC electric power transmission or electric power distribution. PLCs have proven to be a cost-effective solution in a large number of scenarios. Moreover, it provides a distribution system operator with a proprietary communication network and innately integrates the sensing and communication functionalities. Consequently, it has become the predominant smart metering technology in the EU and China [24].

Ethernet is the protocol of choice compared to fiber infrastructure for short and long distances. This technique injects a high-frequency carrier into power lines and modulates the carrier with the data to be transmitted [25]. Typically, Ethernet connections are rated at 1, 10, 40, and 100 Gbps, depending on the technology used [26]. Coaxial cable is a high-speed data transfer technology based on cable television infrastructures. Coaxial cable networks were primarily designed for broadcast services, including television and radio channels. Coaxial cable communication is employed as a communication link between home devices, such as smart meters, an electric distribution company, home automation services, home security, and energy management systems in the smart grid context. Its disadvantage is that the entire bandwidth is shared along the line among many customers making the connection slow [25].

ZigBee [24, 27] is an efficient and cost-effective wireless mesh network built on the IEEE standard 802.15.4. However, it offers a low data rate for personal area networks (PANs). The technology can be employed in device control, reliable messaging, home and building automation, remote monitoring, consumer electronics, health care, and several other areas. Estimated data rates are 250 kbps per channel in the unlicensed 2.4 GHz band, 40 kbps per channel in the 915 MHz band and 20 kbps per channel in the 868 MHz band [28].

Wi-Fi technologies consist of 802.11n (300 Mbps), 802.11b (11 Mbps), 802.11 g (54 Mbps) and 802.11a (54 Mbps) [28]. WI-FI support the computer, laptop, game console or peripheral devices. Wi-Fi is generally an upper layer protocol, with IP being the most predominant protocol, allowing communications over the internet without needing a protocol translator. Smart meters with Wi-Fi modules may be utilized for signal repetition, and the addition of repeaters increases the coverage area and network capacity [28]. Bluetooth [28, 29] is another common wireless communications system used to exchange data over short distances. It employs short-wavelength radio transmission (2400–2480 MHz). Its main features are low power consumption and fast data exchange, and widespread availability. Bluetooth technology can be a viable alternative for the communication of control signs and transmit vitality utilization information.

GSM modem [28, 30] operates in similar ways to the mobile phone because they both require internet connectivity to send and receive information. A GSM modem comprises a dedicated modem device with a USB, serial, or Bluetooth connection. Communication with the GSM can be carried out using machine instructions to activate structures on an intelligent modem known as AT command set. The AT command set is widely known as the Hayes standard AT command set. This functions as a set of instructions for configuring and controlling modems. The commands are short sequences of ASCII characters. All command strings (that is, sequences of characters) must be supplementary by the letters AT, an abbreviation for attention that accounts for the set name.

4. Performance evaluations

4.1 Proposed smart meter block diagram

The smart energy operational block diagram in **Figure 4** depicts the components of making the smart energy meter for an advanced metering system, thus lessening consumers' stress in purchasing energy credit units from vendors' utilities. The device will reduce the production cost, billing cost, and maintenance cost of procuring one from the utility viewpoint.

The smart meter measures the current, voltage, power, and energy consumed by loads. The energy meter comprises the voltage and current sensor that helps with the voltage and current signals' acquisition. The amount of power utilized, the voltage, and current per time are evaluated, enabling the consumer to understand its consumption. More so, energy usage per time is derived per time, thereby providing a fast energy management method. The metering system is also

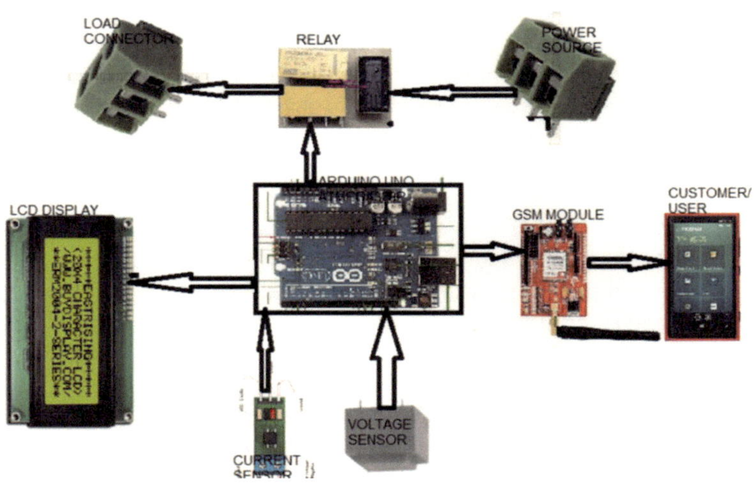

Figure 4.
Smart meter components.

responsible for relaying the amount of voltage and current consumed by the load to the micro controlling unit for the required parameter computation. Hence, if the measured power rating exceeded 2000 Watts, the micro controlling unit sends a command to the relay to control and reduce consumption rate charges. Therefore, the whole system starts to return the entire process to the initialization input all over again. The code in the micro controlling unit is shown in the appendices.

4.2 Performance assessment

The meter was designed with technical specifications that are identified as accuracy (class 1.0); rated voltage; single-phase (230 V → 250 V); frequency (50 Hz/30A); display (LCD), information record, and energy parameters such as power, current, voltage, power, energy, and cost of billing.

The proposed smart meter was simulated using proteus software. Proteus combines mixed mode SPIC circuit simulation and animated components with various microprocessor models, which facilitate simulation. This assists in developing design and test cases. It emerges amongst the simulation software for electronic design.

The simulated design shown in **Figure 5** displays the initialization stage of the smart energy meter. The components are interfaced through the connecting probe. It is seen that the schematic diagram within the simulation showed that the power supply is connected to a potential transformer serving as the voltage sensor. A Zener diode protects the microcontroller unit against any upsurges. The current sensing is based on the Hall effect sensor, with its output increasing by 60 mV for every

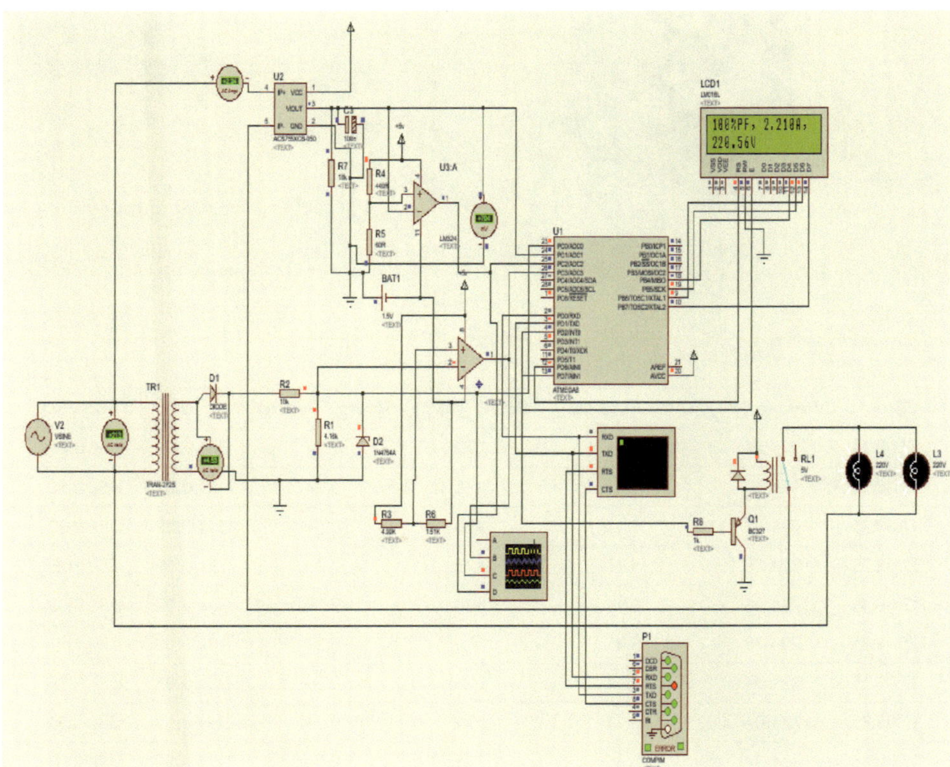

Figure 5.
Smart meter simulated diagram.

ampere increment in the measured current. For the voltage sensor, when no current is flowing in the circuit, the device voltage is 0.6 Volt, which is directly proportional to an increase in voltage when increased linearly by 60 mV/A. Caution is taken to ensure that the measured voltage does not exceed the microcontroller's reference voltage. This is achieved using the zero-crossing detector for enhanced current and voltage measurement.

The zero-crossing detector is a device used for the detection of voltage and current crosses in whichever direction. However, a comparator can be used as a zero-crossing detector. Assuming our reference voltage for the comparator is chosen as zero (Vref =0), the input voltage will saturate the comparator. Therefore, two Op-Amp is employed in place of zero-crossing. Both Op-Amps are configured so that their output goes high whenever their negative input goes lower than zero. The voltage sensor minimum voltage is set to 0.6 Volt.

The circuit has a transistor-driven relay connected to the collector side. The voltage impressed on this relay is a rated full coil voltage at the peak period. Although, in OFF time, the voltage is completely zero to avoid any hazard during use. The PNP transistor is connected to control the switching of the relay. This process facilitates the selection of BC 327 PNP transistors because of their capacity to handle the current, voltage, and power supply. The transistor is also driven into

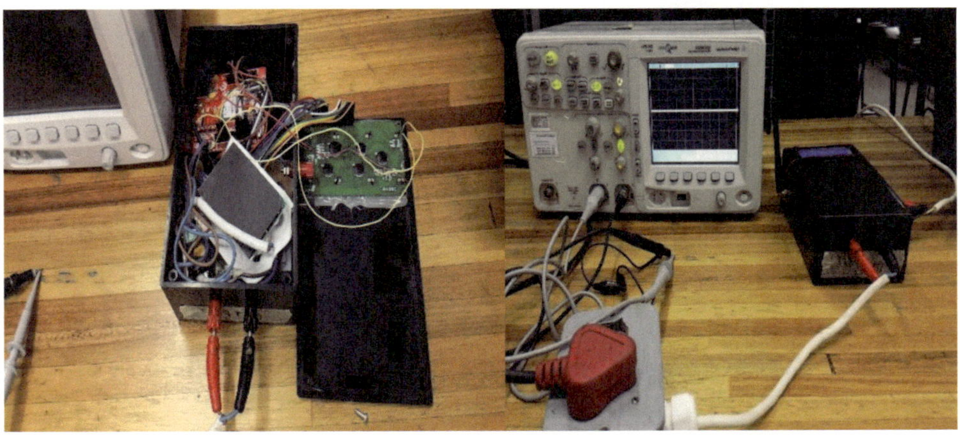

Figure 6.
Designed smart meter.

Time	Voltage (V)	Current (AMP)	Power (kW)	Energy (kWh)	Cost	Total cost
17:08:34	224:51	5.41	1.22	0.34	US$ 0.08	US$ 0.08
17:08:35	224.28	5.41	1.21	0.34	US$ 0.08	US$ 0.16
17:08:35	224.28	5.41	1.21	0.34	US$ 0.08	US$ 0.23
17:08:35	224.28	5.47	1.23	0.34	US$ 0.08	US$ 0.3
17:08:35	224.28	5.44	1.20	0.34	US$ 0.08	US$ 0.38
17:08:35	224.28	5.41	1.21	0.34	US$ 0.08	US$ 0.5
17:08:35	224.04	5.44	1.22	0.34	US$ 0.08	US$ 0.54
17:08:35	224.04	5.44	1.22	0.34	US$ 0.08	US$ 0.61
17:08:35	224.04	5.39	1.21	0.34	US$ 0.08	US$ 0.69

Table 1.
Result of smart energy meter when loaded with fan and air blower.

saturation (turned ON) when the Logic 1 signal is written on the port pin. Thus, turning ON the relay. The relay is turned OFF by writing Logic 0 on the Port 5 and 13 of the ATmega328P. Also, a free-wheeling diode 1 N4148 is connected across the relay coil. This is done to protect the transistor from damage due to the back electromotive force (EMF) generated within the relay's inductive coil. Thus, the transistor is turned OFF. The energy is stored in the inductor as dissipated through the diode and the relay coil's internal resistance when the transistor is switched OFF.

The designed smart meter is depicted in **Figure 6**, while its tested results are tabularized in **Table 1**, based on the meter's response when a fan and a blower are connected. The results show the voltage, current, power, energy, the resulting cost of energy every second, and the cumulative cost of energy.

5. Cost assessment

Table 2 presents lists of all variables considered in the smart energy meter design and development, including their costs. The overall cost of the designed smart meter prototype was evaluated to approximately US $ 157. The cost of producing a unit may seem expensive due to the procedures and methods of executing the design. However, a cost comparison between the developed smart energy meter

S/N	Component name	Manufacturer	Pieces	Cost (US $)
1	USB TTL Serial/RS232 Converter	EIE	1	6.16
2	Term N/C PCB 2 W 2.54 GRN	DEGSON	4	1.07
3	ENCL ABS N/R BK 197 x 114 x 62	Plaster Converter	1	8.14
4	Socket Banana 4 mm 6A w/h Red	EIE	2	1.35
5	Socket Banana 4 mm 6A w/h Black	EIE	2	1.39
6	Plug Banana 4 mm Stack Rub BLK	ELE	2	2.51
7	Plug Banana 4 mm Stack Rub Red	EIE	2	2.47
8	PSU W/M I-90/264 o = 09 V @2A2	HG POWER	1	19.71
9	TRF P = 220 S = 9.5 V 1.5A PCB	EIE	3	9.64
10	Zener DO-35 500 mW 5.1 V 1N5231B	Fairchild	12	0.22
11	Terminal Block PCB 10 mm 2 W SIL	DEGSON	2	0.70
12	Current Detector Board	EIE	1	5.57
13	CAP ELEC RAD 1000uf 6 V3	RUBYCON/HITANO	4	1.86
14	PS TO92 EBC 50 V 0.8A 60 M 160	SOT TECH	2	0.07
15	PS TO92 EBC 50 V 0.8A 60 M 160 SMD	NXP	2	0.04
16	Header SIL STR 40 W 2.54	GTX	1	0.225
17	Jumper Wires	ARD117E (40 15 cm)	1	6.78
18	GSM Shield SIM900	KEYES	1	68.64
19	ARDUINO UNO R3	CPUT	2	0.04
20	LCD1602 module(16x2)	HD44780 Adafruit	2	12.92
US$ 156.93				

Table 2.
Smart energy meter individual component costs.

prototype and selected intelligent energy meters (See **Table 3**) with similar functionalities available in the market was conducted. This comparison demonstrated that the project is cost-effective. For mass production on a commercial scale, the cost will further reduce since components are purchased in bulk.

	Cost	Available smart energy meter	Cost
Designed low cost smart energy meter	US$ 156.93	SMA energy meter	US$ 428.57
		CAK smart metering	US$ 142.86
		DMED 130 meter	US$ 176
		Linky rollout	US$ 186
		Ontec systems/Itron SA	US$ 103

Table 3.
Smart energy meters in the market.

Furthermore, economies of scale can be described as the cost benefits companies acquire when production becomes effective. It is of utmost importance for every company to increase its production, enhancing the lowering of costs. Reference [31] states that mass production and mass customization determine manufacturers' products' behavior. A system that engages mass production operates within a standard that generally accepts and forecasts price reduction through economies of scale. And the price difference between mass-produced and customized goods helps lower the prices of units to achieve 'low-cost' in its generality.

6. Conclusions

The chapter presents a smart energy meter design that meets low-cost, energy-efficient, robust, and multi-functional requirements. The device was developed to measure energy consumption rates and billing. Additionally, the proposed system has added features that allow the recovery of the meter energy measurement data remotely. The system enables monitoring and transmission of energy consumed in real-time. A microcontroller board is used as the controlling unit to execute control and monitor activities. An LCD displays standard electrical measurements such as current, voltage, power, and energy consumption. The external communication device was required in the unit's actualization, in conjunction with the control unit based on the existing mobile technology. It stands as the intermediary between the nearby available utility station and consumers or end-users. In conclusion, liquid crystal display displays real-time based data for the end-user to visualize. The usage data billing is done within thirty seconds, stored, and trans-received the process for data collection, keeping, and billing generation.

Conflict of interest

The authors declare no conflict of interest.

Appendices

Algorithm Used to Program the Microcontroller Unit.

The code below was programmed into the micro controlling unit, debugged, and simulated through proteus with prototype executed in detail.

```
#include <mega8.h>
#include <delay.h>
#include <math.h>
#include <stdlib.h>
#include <string.h>

#include <io.h>
//#include <util/delay.h>

//#include <lcd.h>
//#include "lib/sim300/sim300.h"
//#include <sim300.h>

char *number = "9999999999";
float old_energy = 0;
float reference = 300.0;

//LCD
#define RS PORTD.6
#define E PORTD.7
char t1,z1;

//Global Variables initialization
unsigned char buf[10];

//Power facotr values and functions initialization
void pf_func();
unsigned int k=0,x=0,g=0;
float P=0;
float pf=0;
//________Amperemeter & Voltmeter Functions & Variables start_________

int adc_read(int ch);
int adc;
unsigned char buf[10];
float am=0,energy=0;
float vm=0;

// initialize adc
void adc_init()
{
    // Internal Reference Voltage 2.56
    ADMUX = (1<<REFS0) | (1<<REFS1);

    // ADC Enable and prescaler of 128
    // 8000000/128 = 62500
    ADCSRA = (1<<ADEN)|(1<<ADPS2)|(1<<ADPS1)|(1<<ADPS0);
}

 // read adc value

int adc_read(int ch)
{
    // select the corresponding channel 0~7
    // ANDing with '7' will always keep the value
    // of 'ch' between 0 and 7
ch &= 0b00000111; // AND operation with 7
    ADMUX = (ADMUX & 0xF8)|ch; // clears the bottom 3 bits before ORing
```

```
    // start single conversion
    // write '1' to ADSC
    ADCSRA |= (1<<ADSC);

    // wait for the conversion to complete
    // ADSC becomes '0' again
    // till then, run loop continuously
  while(ADCSRA & (1<<ADSC));

    return (ADCW);
}
//_____Voltmeter Functions End__________
//_____________UART Functions__________
void uart_transmit (unsigned char data)
{
    while (!( UCSRA & (1<<UDRE)));
    // wait while register is free
    UDR = data;
    // load data in the register
}
void string_transmit(char *str){
    unsigned char i=0;
    while (str[i]!=0)
    {
  uart_transmit (str[i]);
  i++;
    }
}

//________Power Factor start______
int powerfactor()
{

k=0;
// To complete number of counts
g=g+1;
//To convert into seconds
pf=(float)g/1000000;

//To convert into radians
pf=pf*50*360*(3.14/180);

pf = cos(pf);
k=abs(ceil(pf*100));

return k;

}
//....................End Power Factor Code
//..........................Begin LCD Code....................................................

int lcd_data(char t)
{RS=1;
PORTB=t;
E=1;
delay_ms(1);
E=0;
```

```
delay_ms(1);
t1 = t << 4;
PORTB=t1;
E=1;
delay_ms(1);
E=0;
delay_ms(1);
return 0;}

int writecmd(char z)
{RS=0;
PORTB=z;
E=1;
delay_ms(1);
E=0;
delay_ms(1);
z1 = z << 4;
PORTB=z1;
E=1;
delay_ms(1);
E=0;
delay_ms(1);
return 0;}

void lcd_print(char *str)
{unsigned char i=0;
while (str[i]!=0)
{lcd_data(str[i]);
i++;}}

void lcd_init(void)
{writecmd(0x02);
writecmd(0x28);
writecmd(0x0c);
writecmd(0x01);
writecmd(0x06);}

void lcd_gotoxy(unsigned char x, unsigned char y)

{
    unsigned char firstcharadrs[] = {0x80, 0xC0,0x94,0xD4};
writecmd(firstcharadrs[y-1] + x - 1);
delay_us(100);

}
//___________End of LCD code___________

/*
interrupt [USART_RXC] void intrp()
{

data1=string_receive1();
while(1){
if(strncmp(data1,"off",3)==0){
PORTD.2=1;}
```

```
lcd_print("House Disconnected");
data1=string_receive1();
if(strncmp(data1,"on",2)==0){
PORTD.2=0;
break;}
 }
}
   */
void Tx_data(char *str)
{
string_transmit("AT+CMGS=");
uart_transmit("");
string_transmit(number);

 uart_transmit("");
uart_transmit('\r');

 while(*str)
{
uart_transmit(*str);
str++;
delay_ms(0);
}

 uart_transmit('\r');
uart_transmit(0x1a);
}

void main(void)
{

//Local Variables
int adc_int[41];
int max=0;
int i=0;
int a = 0;
float max_power = 4000;

// Input/Output Ports initialization
DDRB = 0xff;
DDRC = 0x00;
DDRD = 0b11001100;
//____________UART Initializaion____________
UBRRH=0x00;
UBRRL=12;
UCSRA=(0<<RXC) | (0<<TXC) | (0<<UDRE) | (0<<FE) | (0<<DOR) |
(0<<UPE) | (1<<U2X) | (0<<MPCM);

UCSRB=(1<<RXCIE) | (0<<TXCIE) | (0<<UDRIE) | (1<<RXEN) |
(1<<TXEN) | (0<<UCSZ2) | (0<<RXB8) | (0<<TXB8);
UCSRC=(1<<URSEL) | (0<<UMSEL) | (0<<UPM1) | (0<<UPM0) |
(0<<USBS) | (1<<UCSZ1) | (1<<UCSZ0) | (0<<UCPOL);
#asm("sei")

//__LCD Initialization____
lcd_init();
lcd_gotoxy(1,1);
```

```
lcd_print("SE METER");
string_transmit("SE METER\r\n");
while(1)
{
 if (a == 0){ PORTD.3 = 1; a = 1;} // Pin n goes high
else{ PORTD.3 = 0; a = 0;} // Pin n goes low; // (PORTD.3 == 1

UCSRB=(1<<RXCIE) | (0<<TXCIE) | (0<<UDRIE) | (1<<RXEN) |
(1<<TXEN) | (0<<UCSZ2) | (0<<RXB8) | (0<<TXB8);

delay_ms(1500);
//_____POWER FACTOR____
pf_func();
x = powerfactor();

P=x;
delay_us(20);
lcd_init();
itoa (x,buf);
lcd_print(buf);
lcd_data('%');
lcd_print("PF");
lcd_data(',');
lcd_data(' ');

//______Ammeter_____
// Initialize ADC
adc_init();

for(i=0; i<=40; i++)
        {

 adc_int[i] = adc_read(1); // read adc value at PORTC.1
        }

        max=adc_int[0];
for(i=0; i<=40; i++)
        {

        if(max<adc_int[i])
        max=adc_int[i];

        }

adc=max - 240;
itoa(max,buf);
//am = (float)(adc*0.006849);// 7/1024
am = (float)(adc*0.0416709 *0.7071);// 32.67/(1024 - 240)
ftoa(am,3, buf);
lcd_print(buf);
lcd_data('A');
lcd_data(',');

//______________Voltmeter___________

 adc_init();
for( i=0; i<=40; i++)
        {
adc_int[i] = adc_read(0); // read adc value at PORTC.0
```

```
        }
        max=adc_int[0];
for( i=0; i<=40; i++)
        {
        if(max<adc_int[i])
        max=adc_int[i];
        }

adc=max;
itoa(max,buf);
vm = adc*0.30585 * 0.707; //313/1024
ftoa(vm,2, buf);

 lcd_gotoxy(1,2);
lcd_print(buf);
lcd_data('V');
delay_ms(700);
lcd_init();
lcd_print("***POWER***");
lcd_gotoxy(1,2);

P=P/100;
am=am*vm*P;
if (am/P > max_power){
PORTD.2=1;
}
if (am/P < max_power){
PORTD.2=0;
}

ftoa(am,2, buf);
      //string_transmit(buf);
      // string_transmit("\n\r");
lcd_gotoxy(1,2);
lcd_print(buf);
lcd_data('W');
delay_ms(700);
      am=3.4*am;
      am=am/3600;
      energy=am+energy;
ftoa(energy,2, buf);
      //string_transmit(buf);
      //string_transmit("\n\r");
lcd_init();
lcd_print("***ENERGY***");
lcd_gotoxy(1,2);
lcd_print(buf);
lcd_print("Wh");

      if ((int)energy > (old_energy + reference)){
old_energy = (int)energy;
ftoa(old_energy,2, buf);
Tx_data(buf);
Tx_data("KWH\n\r"); }
}
```

```
}
//________POWER FACTOR FUNCTIONS___________
void pf_func()
{
while(1)
{
if ( PINC.2==1 )
{
TCNT1=0;
TCCR1B = 0x01; // Start timer at Fcpu/1
break;
}
else
 {
 continue;
}
}
while(1)
{
if ( PINC.3 == 1 )
{
TCCR1B = 0x00;
g=TCNT1;
break;
}
else
 {
 continue;
}
}
}
```

Author details

Akindele Olufemi Abegunde, Doudou Nanitamo Luta* and Atanda Kamoru Raji
Cape Peninsula University of Technology Bellville, Western Cape, South Africa

*Address all correspondence to: 212123254@mycput.ac.za

References

[1] Effah E, Owusu KB. Evolution and Efficiencies of Energy Metering Technologies in Ghana. Glob J Res Eng Electr Electron Eng [Internet]. 2014;14 (4–5):1–9. Available from: https://globaljournals.org/GJRE_Volume14/4-Evolution-and-Efficiencies.pdf

[2] Sun Q, Li H, Ma Z, Wang C, Campillo J, Zhang Q, et al. A Comprehensive Review of Smart Energy Meters in Intelligent Energy Networks. IEEE Internet Things J [Internet]. 2016;3(4):464–479. Available from: https://ieeexplore.ieee.org/document/7365417

[3] Mohassel RR, Fung AS, Mohammadi F, Raahemifar K. A survey on advanced metering infrastructure and its application in Smart Grids. In: Canadian Conference on Electrical and Computer Engineering [Internet]. Toronto, ON: IEEE; 2014. p. 1–8. Available from: https://ieeexplore.ieee.org/stamp/stamp.jsp?tp=&arnumber=6901102

[4] Zivic NS, Ur-Rehman O, Ruland C. Evolution of smart metering systems. In: 23rd Telecommunications Forum, TELFOR 2015 [Internet]. Belgrade: IEEE; 2015. p. 635–638. Available from: https://ieeexplore.ieee.org/stamp/stamp.jsp?arnumber=7377547

[5] Phair N. Smart Meters: what does a connected house really mean? Cent Internet Saf [Internet]. 2017;(April):1–11. Available from: https://stopsmartmetersau.files.wordpress.com/2019/10/smart-meters_final.pdf

[6] Bhagawati M, Hiremani V, M S. Smart Energy Meter Management using GSM. Ijireeice [Internet]. 2017;5(2): 7–11. Available from: https://ijireeice.com/wp-content/uploads/2017/04/IJIREEICE-NCAEE-2.pdf

[7] EEI and AEIC. Smart Meters and Smart Meter Systems : A Metering Industry Perspective [Internet]. 2011. Available from: https://www.cpsenergy.com/content/dam/corporate/en/Documents/SmartGrid/A Metering Industry Perspective.pdf

[8] Babu Babel. What is so smart about smart meters? 2015;1–9.

[9] Weranga KS., Kumarawadu S, Chandima D. . Smart Metering Design and Applications. Springer Berlin Heidelberg; 2014. 1–146 p.

[10] Depuru SSSR, Wang L, Devabhaktuni V. Smart meters for power grid: Challenges, issues, advantages and status. Renew Sustain Energy Rev [Internet]. 2011;15(6):2736–2742. Available from: http://dx.doi.org/10.1016/j.rser.2011.02.039

[11] Halder T. A Smart metering system. In: 1st International Conference on Non Conventional Energy (ICONCE 2014) [Internet]. Kalyani: IEEE; 2014. p. 204–209. Available from: https://ieeexplore.ieee.org/document/6808720

[12] Andreadou N, Guardiola MO, Fulli G. Telecommunication Technologies for Smart Grid Projects with Focus on Smart Metering Applications. Energies [Internet]. 2016; 9(2):1–8. Available from: doi:10.3390/en9050375

[13] Rout J, Scotty J. Outsmarting your smart meter: A guide to smart meters and how you can use them for cheaper electricity bills. Bundl.; 2009. 28 p.

[14] Himawan H, Supriyanto C, Thamrin A. Design of Prepaid Energy Meter based on PROTEUS. In: 2nd International Conference on Information Technology, Computer, and Electrical Engineering (ICITACEE) [Internet]. Semarang: IEEE; 2015. p. 243–247. Available from: https://www.semanticscholar.org/paper/

Design-of-Prepaid-Energy-Meter-based-on-PROTEUS-Himawan-Supriyanto/153d5ded7e64df9498626ce9ab000ae8f1362ac6

[15] Ahmad J, Malik TN, Awan BA. Design and Study of a GSM Smart Energy Meter to Reduce Load-shedding and Cost of Energy. In: International Conference on challenges in IT, Engineering and Technology (ICCIET'2014) [Internet]. Phuket; 2014. Available from: https://dx.doi.org/10.15242/IIE.E0714003

[16] Effah Emmanuel, Owusu KB. Evolution and Efficiencies of Energy Metering Technologies in Ghana. Glob J Res Eng Electr Electron Eng. 2014;14(4–5):1–9.

[17] Islam KT, Islam AMJ, Pidim SRH, Haque A, Khan TH, Morsalin S. A smart metering system for wireless power measurement with mobile application. Proc 9th Int Conf Electr Comput Eng ICECE 2016. 2017;131–4.

[18] Youssef TA, Elsayed AT, Mohammed OA. Data distribution service-based interoperability framework for smart grid testbed infrastructure. Energies [Internet]. 2016;9(3). Available from: https://www.mdpi.com/1996-1073/9/3/150/pdf

[19] Abdel-majeed A, Tenbohlen S, Ellerbrock A. Design and Building of a Cheap Smart Meter. In: IEEE Power and Energy Student Summit. Bielefeld: IEEE; 2013. p. 1–3.

[20] Maitra S. Smart Energy meter using Power Factor Meter and Instrument Transformer. Commun Appl Electron [Internet]. 2016;4(1):31–37. Available from: https://citeseerx.ist.psu.edu/viewdoc/download?doi=10.1.1.740.5382&rep=rep1&type=pdf

[21] Ajenikoko GA, Olaomi AA. Hardware Design of a Smart Meter. Int J Eng Res Appl [Internet]. 2014;4(9):115–119. Available from: https://core.ac.uk/download/pdf/25574396.pdf

[22] Umang P, Mitul MP. A Review on Smart Meter System. Int J Innov Res Electr Electron Instrum Control Eng [Internet]. 2015;3(12):70–3. Available from: https://www.ijireeice.com/upload/2015/december-15/IJIREEICE15.pdf

[23] Mehta K V, Prajapati B, Wani US. Advance featuring smart energy meter with bi-directional communication 1. In: International Conference on Electrical Electronics and Computer Science [Internet]. 2014. p. 169–74. Available from: https://docplayer.net/43605566-Advance-featuring-smart-energy-meter-with-bi-directional-communication.html

[24] Cortés JA, Idiago JM. Smart Metering Systems Based on Power Line Communications. In: Smart Grids and Their Communication Systems [Internet]. Springer Singapore; 2018. p. 121–170. Available from: http://dx.doi.org/10.1007/978-981-13-1768-2_4

[25] Kuzlu M, Pipattanasomporn M. Assessment of communication technologies and network requirements for different smart grid applications. In: 2013 IEEE PES Innovative Smart Grid Technologies Conference, ISGT 2013 [Internet]. Washington, DC: IEEE; 2013. Available from: https://ieeexplore.ieee.org/document/6497873

[26] Budka KC, Deshpande JG, Thottan M. Communication Networks for Smart Grids: Making Smart Grid Real. Springer-Verlag. Springer Berlin Heidelberg; 2014.

[27] Peng C, Qian K. Development and Application of a ZigBee-Based Building Energy Monitoring and Control System. Sci World J [Internet]. 2014;2014. Available from: https://downloads.hindawi.com/journals/tswj/2014/528410.pdf

[28] Mahmood A, Javaid N, Razzaq S. A review of wireless communications for smart grid. Renew Sustain Energy Rev [Internet]. 2015;41:248–260. Available from: http://dx.doi.org/10.1016/j.rser.2014.08.036

[29] Uribe-Pérez N, Hernández L, de la Vega D, Angulo I. State of the Art and Trends Review of Smart Metering in Electricity Grids. Appl Sci [Internet]. 2016;6(3):1–24. Available from: https://www.mdpi.com/2076-3417/6/3/68/pdf

[30] Sultan Z, Jiang Y, Malik A, Ahmed SF. GSM based smart wireless controlled digital energy meter. In: ICETAS 2019–2019 6th IEEE International Conference on Engineering, Technologies and Applied Sciences [Internet]. Kuala Lumpur: IEEE; 2019. p. 3–9. Available from: https://ieeexplore.ieee.org/document/9117479

[31] Radder L, Louw L. Mass customization and mass production. TQM Mag. 1999;11(1).

Chapter 2

Advancing Off-Grid Electrification by Uncovering the Holistic Risk Landscape Using a Standardized Risk Management Procedure (SRMP)

Inken Hoeck and Elmar Steurer

Abstract

Although it is now well known that access to electrification is a crucial prerequisite for ensuring sustainable development, rural households in sub-Saharan Africa in particular remain unelectrified. It is often not economically viable to connect these remote communities to the main grid. Therefore, mini-grid systems represent a promising alternative to ensure electrification even at long distances from the grid, backed by the fact that these systems are becoming cheaper with the advancement of integrated technologies. However, such systems are fraught with risk if various potential pitfalls are not considered upfront. This discourages investors and thus prevents the electrification rate to increase. The following chapter therefore aims to highlight the risk landscape for the deployment of mini-grid systems in order to assist investors in sustainably integrating mini-grid systems. The approach is illustrated using Namibia's largest mini-grid in Tsumkwe as an example. Through the application of the SRMP, it is revealed that the mini-grid is exposed to a moderate level of risk, mainly due to a lack of education and a replacement process.

Keywords: Rural Electrification, Risk Assessment, Off-Grid System, Mini-Grid System, Namibia

1. Introduction

Despite the fact that energy is undisputedly vital to generate economic growth and ensure a common social well-being, more than 780 million people worldwide still live without access to electricity [1]. Senegal's GDP increased by 1.7% when a 70 MW power plant was put into operation. The same was observed in Uganda, where GDP increased by 2.6% with the commissioning of a 250 MW hydropower plant [2]. However, sustainable development is stagnating in those regions without electricity access. This is particularly true for populations located in rural areas that are difficult to reach, especially in sub-Saharan Africa (SSA) [3]. While the access rate in SSA rose in the past years especially due to the achievements of a very few countries, the Covid-19 pandemic confound this development and added urgency to this topic [4].

Off-grid solar systems provide a feasible solution to electrify these remote areas by closing the access gap and offering reasonable costs and shorter waiting times compared to grid expansion [4, 5]. Apart from a sound technological concept, several other (risk) dimensions influence the reliability and long-lasting persistence of a mini-grid. This publication exemplifies the holistic risk landscape through the case of Tsumkwe, a settlement in the northeast of Namibia. For this purpose, a risk-rating model for mini-grids, based on the standardized risk management procedure (SRMP), is used. Steurer et al. (2017) provide a more detailed description of this model, which was refined in the course of this paper [6]. The risk evaluation comprises five categories: Regulation, Economy, Technology, Finance and Education, thus providing an indication of the multi-dimensionality of this subject. Being located in an economically disadvantaged rural area of Namibia, Tsumkwe relies on electricity from a mini-grid, which was set up in 2011 and is, therefore, well suited for the topical investigation.

2. The SRMP approach

2.1 The need for a holistic risk assessment

Mini-grids are exposed to various types of risks. It is therefore not unexpected that many of these projects have failed along the way. An aggravating factor certainly is the amount of different stakeholders that are involved in such projects with each following their own objective – electricity distributor, public and private investors, government and eventually the consumer or an enterprise. While the power supplier is usually striving to increase its profits, the government is committed to a just energy supply and the consumer, in turn, demands a relatively cheap and reliable energy supply. This altogether needs to be translated into a reasonable business story, comprising, for example, questions related to the overall organization, education, as well as the financial and economic feasibility (see **Table 1**). This results into a multi-dimensional and complex endeavor, which has to be planned in a thoughtful way. Due to this complexity, a holistic risk assessment tailored to the needs of mini-grids has to be in place. This is achieved with the standardized risk management procedure (SRMP) for mini-grids, which is considered a measure of the ability of a mini-grid to meet financial obligations and profitability expectations in the medium and long term.

Cluster	Question
Regulatory	Is the political situation stable?
	Is the mayor of the respective village accepting and supporting the development of a mini-grid system?
	Has a grid arrival policy been adopted?
	Are further information existent that relate to off-grid systems?
	What is the crime rate?
Economic	How does the tariff design look like?
	How is the billing process structured?

Cluster	Question
	How many commercial customers are connected to the mini-grid (in relation to connected households)?
	Can money be saved for potential repairs or extensions?
Technology	Who is taking care of maintenance and potential repairs?
	Is the used technology easy to maintain?
	Who is taking care of the operation management?
	What does the procurement process looks like?
Finance	Who is responsible for cost/result controlling?
	Who is responsible for Accounting and Budgeting?
	How is the reporting structure designed?
	How does the funding structure look like?
	Where to ask for funding?
Education	Who is responsible for training the (technical) staff? Is a training plan in place?
	Who is responsible for raising awareness and training the (rural) population?
	How can an increase of productive use cases be achieved?
	Is there a place people can approach for help/assistance?
	Are there funding opportunities for small business startups?

Table 1.
Uncertainties related to the five dimensions.

2.2 The structure of the SRMP approach

The SRMP, in general, weights factors related to the regulatory framework, economy, technology, finance and eventually education. These five categories are evaluated based on qualitative and quantitative data, resulting into a risk score that ranges from very low (1) to very high (6). The final score indicates the holistic risk level for implementing mini-grids in a specific region (here: Tsumkwe). Those marks than flow into a combined score (see **Figure 1**).

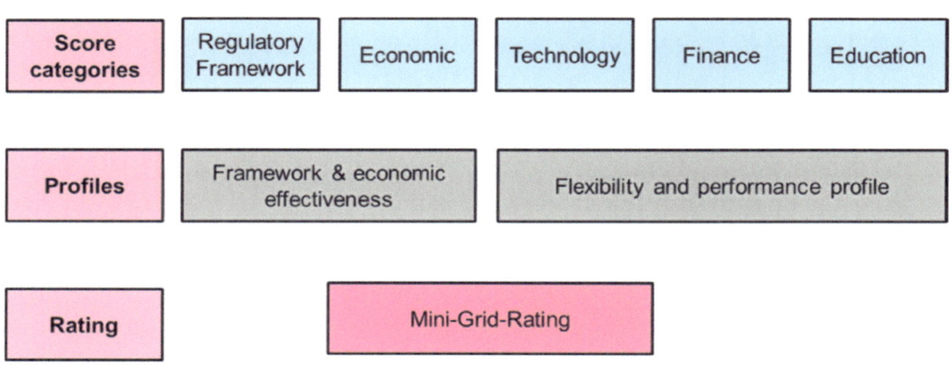

Figure 1.
Framework of the SRMP [6].

Each of the five criteria described previously is weighted individually to obtain a single score of the five criteria to form a profile. For example, the regulatory framework score is combined with the economic score at an equal weighting of 50%

to generate the overall "framework and economic effectiveness" profile. The scores from the remaining three criteria, i.e., the Technology, Finance, and Education scores, each account for 33% of the total "Flexibility and Performance Profile." The two scores from these intermediate profiles are then combined resulting in a rating level that represents the final, overall mini-grid rating (see **Figure 1**).

The overall rating of the mini-network is aggregated and expressed in the form of rating levels from AAA to D and is derived from an equal weighting of the intermediate profiles at 50% each. Consequently, the ratings obtained correspond to the credit ratings commonly used on the capital markets and published by the international rating agencies, e.g. Standard & Poor or Moody's and Fitch. The reason for this is to provide institutional investors with a decision-making tool with which they are familiar, one of the stakeholders in mini-grid projects.

3. Electrification in Namibia

The proportion of the Namibian population with access to electricity grew from 43.7% in 2010 to 57.4% in 2019 [4]. This might be surprising, as Namibia is frequently perceived as developed as South Africa when it comes to the electrification rate. Namibia, however, is in a somewhat more difficult situation, as being the second least densely populated country in the world [7]. The population is widely scattered, which makes it almost impossible to connect all villages to the main grid. Especially remote areas are unable to use electricity, which hinders the development. Thus, off-grid systems represent a great opportunity to enhance rural electrification. In particular, mini-grids based on solar energy provide great value in view of the potential of solar energy in Namibia, as the country exhibits the second highest level of solar irradiation in the world [8]. In addition, Namibia is today heavily dependent on fossil fuel imports from neighboring countries such as South Africa [9, 10]. Encouraging current efforts to increase the penetration of solar energy supply is therefore an essential step not only toward greater energy independence, but also toward lower national electricity prices and an overall reduction in CO2 emissions.

In the Vision 2030, issued in 2004, the Namibian government has already stated to pursue a sustainable energy policy in order to accelerate urban and rural development [11]. The same was declared in the Strategic Plan 2017/2018–2021/2022 devised by the Ministry of Mines and Energy (MME) [12]. This indeed appears to be a valid strategy, considering the above-mentioned predicament.

3.1 The electricity sector in Namibia

The Electricity Supply Industry in Namibia has started a transformation process in the 2000s. Prior to that, the state-owned national power utility 'NamPower' had a quasi-monopoly in the market, being responsible for the generation, transmission and distribution of electricity. While NamPower was the only entity authorized to supply electricity to farms and mines, Local Authorities (LA) and Regional Councils (RC) were exclusively liable for providing electricity to residents and businesses. [7] Starting in 2002, regional electricity distributors (REDs) have been conquering the market, resulting in progressive liberalization. This development also created a market for private Independent Power Producers (IPPs). However, sales to end customers continue to be handled exclusively by NamPower, the REDs and municipal utilities [7].

In 2000, the Electricity Control Board (ECB) took over the role of market regulator, which is defined in the Electricity Act of 2007 (formerly: Electricity Act of 2000). In this function, the ECB issues regulations and technical standards for the connection of renewable energy (RE) plants to the grid, while the Ministry of Mines and Energy (MME) is in charge of the development of the industry itself and ultimately acts as a policy maker [12].

3.2 State of electrification in Tsumkwe

Tsumkwe is located in the northeastern part of the Otjozondjupa Region in Namibia. The settlement is 735 km away from Windhoek and 304 km from Grootfontein, the nearest town where community members have access to basic services (e.g. banks or supermarkets). According to an observation trip conducted in 2005, the closest grid electricity access point from Tsumkwe is either in Maroelaboom, which is about 180 km to the west (33 kV connection) or at the Berg Aukas Distribution Station located about 240 km to the west (66 kV connection) [13]. Based on Namibia's Rural Electricity Distribution Master Plan (REDMP), Tsumkwe is considered an off-grid area and will not be connected to the grid in the near future [14]. This is mainly due to the high cost of N$150 million associated with the connection [15].

Prior to Namibia's independence, Tsumkwe served as a military base for the South African Army. The power supply was provided by means of two diesel generators connected to a micro-power grid with a small medium voltage of 11 kV [15, 16]. After Namibia's independence in 1990, the government funded the construction of a school, a clinic and a police station for the community, which could be utilized through the infrastructure left by the South African government [17]. However, power supply was unreliable and due to increasing costs of diesel, electricity supply started to be restricted, which inhibited the dissemination of businesses and had an adverse influence on livelihood in general. As the system generally was poorly maintained, diesel occasionally spilled and polluted the environment. The tariff was not affordable for residents until that time, although the government heavily subsidized it, leading to a tariff of 0.14 USD/kWh for private households and 0.27 USD/kWh for institutions and commercial customers (equivalent to 2 and 4 NAD respectively based on the exchange rate of 2014). The subsidies, furthermore, resulted into an annual deficit of NAD 1.2 million for the Otjozondjupa Regional Council (OTRC) [16].

Eventually in 2005, the Councilor of Tsumkwe called for an improvement of the electricity situation. Therefore, a small team of experts was commissioned by the MME through the Desert Research Foundation of Namibia (DRFN) to evaluate the current energy situation in Tsumkwe. The team was precisely instructed to assess whether a hybrid mini-grid energy supply system using solar energy and diesel would be a feasible long-term electrification approach for Tsumkwe. Based on good experience with a project at the Gobabeb Training and Research Centre, which ensured a reliable electricity supply through a solar PV system backed by a diesel generator, it was decided to use the same hybrid approach for Tsumkwe. The usage of solar power is not at all surprising, as Namibia exhibits the second highest level of solar irradiation in the world [8].

The European Commission, NamPower and the OTRC, funded the resulting project. The Namibian government's share thereby primarily benefited the introduction of energy efficiency appliances. For example, solar water heaters were substituted for electric water heaters, electric stoves were replaced with gas burners, and households were equipped with energy-efficient light bulbs. All this was done on the basis of the recommendations of the experts from the observation tour in 2005. Although the experts' advice was additionally to run the hybrid mini-grid system autonomously through an independent operator, who could collect revenue and take responsibility for maintaining the system and ultimately developing the village, ownership of the system was transferred to the OTRC. Operation and maintenance of the mini-grids was under the responsibility of the Department of Works residing in Tsumkwe [13, 15].

4. Risk assessment on the example of Tsumkwe

4.1 Regulatory assessment

> This criterion mirrors the degree of impact that official bodies and authorities overseeing electrification have on a mini-grid. The transition to preferably RE based off-grid solutions indeed requires strong policy support [3, 9, 18] and as investments in off-grid solutions are largely driven by regulatory policies [5], the importance becomes apparent.
>
> For the sake of comprehensibility, the criterion is divided into two factors. The primary factor assesses the effectiveness, stability, and predictability of political institutions. Particular attention is paid to the alignment of local bodies and the population with a mini-grid project. Therefore, the attitude of local mayors towards the electrification project is rated and the support of the local population is measured. Moreover, the community's crime rate is considered, as this poses the risk of ongoing conflicts with potentially negative impacts on an electrification project. Another key determinant in the evaluation of the regulatory framework is the degree of transparency, accountability, and process reliability of the various institutions involved. Simply put the extent to which local bodies take action to signal their accountability for their policy decisions and their outcomes. This area is captured by the secondary factor. Obvious conflicts of interest are one of the things that have to be taken into account. This requires critical scrutiny when analyzing the organizational structure of the mini-network. Further, generally unbiased enforcement of contracts and respect for the rule of law add to this.

4.1.1 Primary factor

It is essential to accelerate the deployment of solutions that provide Namibian rural areas decentrally with access to RE in order to, ultimately, facilitate the country's development. That mini-grids are acknowledged as a valid option for energy generation by the government is highlighted in the Renewable Energy Policy, outlining "(...) while some regions of Namibia are most optimally served through grid-connected, utility-scale renewables, other locations are well-suited to being powered by off-grid applications. Both approaches are complementary in nature and neither grid extension (with renewables integration) nor off-grid systems by themselves can provide a solution in isolation." [19]. Despite this, the regulatory framework in Namibia that addresses off-grid electrification, the OGEMP, shows a number of deficiencies.

The document fails to provide a plan for a possible grid expansion. Nonetheless, this is an essential piece of information, not least because mini-grids have been abandoned in recent years by the time a village was connected to the national grid [18], which certainly yields uncertainty amongst investors. Thus, policies are required that define rules for the further persistence of mini-grids and preferably ensure investors that mini-grids are retained and continued to be used to sell generated electricity to the grid after the grid is extended [20]. That is already underlined by the first Policy Statement (P1) of the National Energy Policy, which declares to "create opportunities for mini- and micro-generators to feed into the national grid and off-grid mini-grid networks." [12]. Moreover, in the latest version of the OGEMP, the government provides a rough time frame for planned grid connections in certain municipalities. Although this could reduce uncertainty and encourage investments in communities that have no perspective of being connected to the grid (such as Tsumkwe), uncertainty remains in pre-grid areas. Consequently, a comprehensive policy in that regard is yet missing. The OGEMP secondly misses the chance to explicitly present off-grid possibilities, such as mini-grid or stand-alone systems, although including clear and transparent information on the different solutions for off-grid energy generation, potentially reduces barriers to market entry. Lastly, the focus within the OGEMP is almost exclusively placed on an 'energy shop approach'. According to this approach, private households, businesses and institutions gain access to energy technologies through these shops, which are planned to be established in "reasonable distances" – corresponding to 10 km to 30 km [21]. Energy shop owners are also supposed to consult their customers regarding funding, assist by hiring technicians when needed and they are ultimately the ones who do not only receive loan payments, but also pursue missing or late payments [21]. Yet the document does not address a conceivable reluctance of storeowners to sell energy products, opting instead for the "over their heads" strategy. Past observation has demonstrated that owners do not necessarily stock up on energy equipment, in part because of a lack of capital. Moreover, in reality, very few energy stores have sufficient expertise in subsidies and relevant application mechanism [22]. Overall, there is a lack of policies, frameworks or even institutions specifically related to mini-grids, although the country convinces with a very transparent regulatory framework especially in the context of renewable energies [23]. This leads to a final rating of 3 (see **Table 2**).

As for the legal basis for the generation, distribution and sale of electricity, Namibia has already introduced a transparent licensing process, which takes about 60 days to be completed. In comparison, the process takes 90 days in Kenya [24]. As stated in the Namibian Electricity Act, an electricity generation license is not required for projects of 500 kVA or less, which reduces development costs and thus, attracts investors. A streamlined licensing framework for off-grid systems is not yet available, which potentially "reduce[s] the regulatory process involved in obtaining licenses or permits, reducing costs for off-grid operators" [25].

Apart from the national circumstances, other predominant conditions in Tsumkwe are included in the assessment that feed into the primary factor. For example, the relatively low crime rate and the affirmative attitude of the settlement mayor towards the mini-grid systems, which both affect the evaluation positively. The Tsumkwe community was partially involved in the planning and deployment phases of the mini-grid. Local employment for the construction was particularly important to establish a sense of ownership of the infrastructure. In addition,

Primary factor: The effectiveness, stability, and predictability of the policymaking and political institutions on community and block level		Secondary factor: The transparency, accountability and process reliability of the other involved institutions, especially the block and district	
Alignment major and local authorities	2	Obvious conflicts of interests	1
Crime rate in Tsumkwe	2	Generally unbiased enforcement of contracts and respect for the rule of law	2
National Policies	3	Significant and sustained issues between entities on state level and local level	3
Result	2.33		2.0
Final score	2.25		

Table 2.
Regulatory Framework scoring.

educational campaigns were arranged that focused specifically on informing the community about the project itself, how to maintain the stoves/solar water heater, and about energy efficiency. This included several informative flyers that educated residents on how to save money through energy-efficient practices, among other things [13]. Despite these efforts, the community was not engaged in the course of setting tariffs, which however is essential to ensure a sustainable management of the plant. Considering all this, the overall risk score of the primary factor is 2.33 (see **Table 2**).

4.1.2 Secondary factor

Moderately well documented are non-financial instruments, such as technical standards or grid codes for off-grid systems, which supposedly strengthens the emergence of IPPs by providing transparent information on market requirements. There is a continuing lack of guidelines in this context, which are directed not only at private investors, but also at the REDs by means of providing training material on the installation and maintenance of mini-grids. This can be seen in the case of Tsumkwe. During the time the mini-grid was owned by the RC, it has not been operated as designed and not been maintained at all [15]. Once the takeover was planned, NamPower and other REDs have been immensely reluctant to inherit the responsibility for the operation and management of mini-grids, due to the "lack of viability, relevant expertise, and regulatory uncertainty" [22]. As a consequence, the mini-grid is still not effectively taken care of [15, 26]. This reflects the insufficient involvement of the government in promoting and supporting off-grid systems as well as the persistent challenges between entities at the state and local levels.

Overall, the secondary factor performs slightly better. This can be traced back to the exceptional transparency of regulatory processes, apart from what has been described above. Deductions are made for the just moderate respect for the rule of law, experienced in Tsumkwe [15]. This generally leads to a result for secondary factor score of 2.0. Since the primary factor is weighted at 75 percent and the secondary factor at 25 percent, the final score is 2.25, indicating a rather low risk level.

4.2 Economic assessment

The economic score is derived by analyzing the purchasing power of the local population (retail score) and the diversity of the regional economic sectors (commercial score).

Both scores are averaged to generate the final economic score. To obtain the retail score, the per capita income of the locality is utilized. Hence, the average monthly income per person is compared to the corresponding average income level of a comparison group. The comparison group is usually dictated by the region. The variety and diversity of the pillars upon which the local economy relies is another critical determinant of the economic score and is illustrated by the commercial score. An economy that relies on different industries is able to withstand crises in one or more sectors of its economy, making its overall economic performance less volatile. Concentration in solely one sector is consequently perceived as a negative. Accordingly, in the case of rural mini-grids, a high share of agricultural activities results in a negative score. Because of this reasoning, the share of agricultural business is taken into consideration for the overall local economic performance. A higher share results in a higher risk for the mini-grid.

It is safe to say that an economically viable mini-grid operation is currently hardly possible in the case of Tsumkwe. Based on a conducted Levelized Cost of Energy (LCoE) calculation, which is backed by assumptions derived from findings of a research stay in 2019, the total costs for generating electricity in Tsumkwe amount to approximately 2.2 NAD/kWh. Revenues, on the other side, are 2.4 NAD/kWh when assuming a share of 80% private households and 20% commercial customers. This leads to a negative profit margin of 0.2 NAD/kWh, indicating a low profitability for potential investors (see **Table 3**).

The risk assessment itself is based on the two dimension outlined above, which are the income level as well as the economic diversification, which underlines the importance of engaging commercial customers.

Expenses	Income	
OPEX costs range from 2.2 NAD/kWh	Residential customers 80% @ 2 NAD/kWh	Average income 2.4 NAD/kWh
	Commercial customers 20% @ 4 NAD/kWh	
	Profit margin 0.2 NAD/kWh	

Table 3.
Profitability analysis (estimation).

4.2.1 Village income level

The village income level of Tsumkwe lies at around 4880 NAD monthly per household, which was elicited during a research trip in 2020. The land is not fertile enough to engage in subsistence or even commercial farming, thus the largest portion of household in the Otjozondjupa Region report salaries or wages as their main source of income [27]. The income is set in relation with the average income level in Otjozondjupa region as of 8317 NAD per household per month [28]. This ratio as of 58.3% corresponds to a risk sub-score of 3.5 (see **Table 4**).

Position	1	2	3	4	5	6	7	8	9
Average income village / avg. Income peer group	100%	90%	80%	70%	**60%**	50%	40%	30%	20%
Income score	1	2	2,5	3	**3,5**	4	5	5,5	6

Table 4.
Income score.

4.2.2 The importance of commercial customers

The electricity tariff for commercial customers is higher, than for private households. Further, commercial customers, as a standard, consume electricity mainly during the daytime, while the consumption of private households peaks in the morning and in the evening. The mini-grid operator benefits from higher daytime utilization of the solar mini-grid, since providing power directly from PV panels equals zero marginal costs. Batteries represent fixed costs, but are depreciated with each discharge. Diesel generators have high variable costs driven by fuel prices. Thus, for the operator, adding daytime demand reduces the overall levelized cost of electricity. Midday loads, as stated above, are more likely to consist of income-generating activities such as, for example, cold storage, processing of agricultural products or running small workshops. Evening loads are more likely to be residential. Having productive-use customers can, therefore, benefit residential customers, since the midday utilization lowers the overall price of electricity. In other words, the existence of more commercial customers, who buy power at zero marginal cost during the day, will lower the price for the electricity residential customers pay at night. In the case of Tsumkwe, however, the share of residential customers contributes the most to the revenues of the mini-grid. This leads to the assumption that the scarceness of commercial customers (among other factors) potentially causes the situation illustrated in the beginning.

The level of productive use in a local economy is, therefore, a crucial determinant for the economic score within the SRMP. A mini-grid that serves customers in the field of productive use can achieve a stable overall economic performance at lower prices for both commercial and retail customers. As a result, isolated concentration on the residential sector or a high share of household or residential activities lead to a negative assessment. Based on this, the share of residential customers feeds into the assessment – a higher share leads to higher risk for the mini-grid. Besides, the share of anchor customers revenues compared to the total revenues of the mini-grid is used for measuring the commercial strength. Anchor customer are, for example, larger enterprises or public institutions. The utilization profile of an anchor customer is rated as both stable and safe due to its high creditworthiness. A higher share of anchor customers to total commercial customers contributes to a lower risk for the mini-grid (**Table 5**).

		Economic diversification					
		Share of residential customers as % of total revenues (less than)					
		0%	**20%**	**40%**	**60%**	**80%**	**100%**
Share of anchor customers to total commercial customers (less than)	**80%**	2	2	3	3	4	4
	60%	2	3	3	4	4	5
	40%	3	3	4	4	5	5
	20%	3	4	4	5	5	6

Table 5.
Economic diversification.

In the case of Tsumkwe, the anchor customers are the police station, lodge, hospital and schools, contributing roughly 22% of the mini-grid's revenues as commercial customers. Together with the low share of existent commercial customers, an overall commercial score is derived [6]. The relatively high residential usage of 75% and the low share of anchor customers among commercial customers of 22%, combined with a medium average income level in Tsumkwe, ultimately lead to a moderate economic score of 3.75 (**Table 6**).

Components	**Score**
Income level	3,5
Economic diversification	4,0
Final Economic score	3,75

Table 6.
Economic score components.

4.3 Technological assessment

The technology score is intended to provide conclusions concerning the reliability and resource efficiency of the system. Therefore, the maintenance cost and the share of renewable energy are both chosen to measure how much attention is paid to providing a robust and reliable technology frame and to measure the resource efficiency of a mini-grid. The maintenance cost is put in relation to the investment, while the share of renewable energy is calculated by the energy generated annually from renewable sources compared to the total energy generated each year. However, any additional source, especially based on renewable energy, results in all probability in inefficient supply structures and additional sources of technical error. If that is the case, the value would be adjusted.

In 2011, the existing diesel-based mini-grid in Tsumkwe was upgraded with a 202 kW_p solar PV plant and a 1.93 MWh lead-acid battery system [15]. After some years of operation, the battery storage and PV panels were extended to 3.08 MWh and 303 kW_p, respectively, which made it the largest off-grid electricity supply system in Namibia. Besides the open-space PV generator and its inverters, the hybrid mini-grid has a diesel generator set to enable a 24-hour, uninterrupted electricity supply for the settlement. The diesel genset has a rated output of 300 kVA corresponding to a true power of 240 kW. As for the distribution side, the mini-grid in Tsumkwe consists of an 11-kV three-phase distribution grid separated into an essential and a non-essential line. This structure allows the consumers connected to the essential line - that is the water pumping station and the clinic - to be given priority in the event of a power or energy shortage. Zongwe et al. (2017) indicate the average daily electricity consumption in Tsumkwe to be about 2 MWh/d [15], which was confirmed during the first field trip within the research project in 2019. In the same period, PV production amounted to 1,618 kWh/d. The diesel genset contributed some 581 kWh/d of electricity, which corresponds to an average running time of about 2.4 hours per day. Hence, PV reached a share of approx. 69% of the total electricity generation in the mentioned period, which indicates an economically optimized mini-grid design: a renewable energy fraction of 60% was identified as cost-optimal for hybrid mini-grids in a publication of IRENA [29][1]. Overall, the general technical design of the mini-grid conforms very well to the present energy situation in Tsumkwe.

However, the research trips have also shown severe lack of maintenance. For example, the PV modules were covered by a thick layer of dust, significantly reducing the solar yield [26]. One PV module was even damaged because of a stone chip, which caused a hot spot. As such damage is very likely to lead to module failure and thus failure of the entire PV string, the research team had to circumvent the defective module. Moreover, the PV inverters were found to be heavily contaminated with dust, which impairs heat dissipation. The resulting overheating in turn reduces inverter performance and the overall lifetime of the system. [23] Besides these defects, the lead acid battery banks have proven to be the weakest part of the Tsumkwe mini-grid. Five defective battery cells were identified during a research trip. Since these battery cells were already short-circuiting due to electrode sludging and thus endangered the entire battery system, they had to be removed from the battery banks immediately. In early 2021, furthermore, the CENORED operations team detected that an inverter was defective and that the continuity of power supply was being affected. The request for support from CENORED Management to the research project team in Germany provided evidence that no replacement process is implemented.

The SRMP scheme combines the low maintenance level in terms of investment costs of 6% together with a share of 50% renewable energy in an initial score of 4 (see **Table 7**).

---

[1] It is expected that the cost-optimal RE fraction may rise to 90% by 2025 as the costs of storage and control devices continue to fall.

		Maintenance costs / Investment costs			
		15%	10%	5%	0%
Share of RE (% of overall electricity generation)	**100%**	1	1	1	2
	80%	1	1	2	3
	60%	1	2	3	4
	40%	2	3	**4**	5
	20%	3	4	5	5
	10%	**4**	5	5	6
	0%	5	6	6	6

Table 7.
Technology score: Initial score.

Value		**Adjustment**
Procurement process for spare parts in place?	No	+1
Complex technology?	No	0
Adjustment total		+1
Total technology score		5

Table 8.
Technology score: adjustments.

In a second step, the technology score is adjusted if there are special technological risks in place (see **Table 8**). In case of the Tsumkwe mini-grid, the missing procurement process for spare parts leads to a negative adjustment by one score, resulting in a final technology score of 5.

4.4 Financial assessment

> The finance score is measured in two steps. First, two quantitative tables are used to obtain an initial score by applying a scoring grid. The initial score is based on a calculation of the average of the percentage of debt and the cost of debt over the last three years and can be improved in the case of a positive adjustment and worsened by a negative one. In this way, financial flexibility and independence is assessed by the amount of net debt as a percentage of total capital and interest expense as a percentage of revenue. The baseline can be changed by a positive or negative adjustment. The baseline could improve by one grade if a grant is awarded to finance the mini-grid. A grant in all probability affects the flexibility of a mini-grid in a positive way and allows the mini-grid's management to operate without financial pressure from investors for extended periods of distress. On the other hand, this must not lead to moral hazard or encourage development that disregards necessary structural investments. To cover this characteristic, the involvement of private investors is being examined. A private investor can be regarded as extremely relevant in order to avoid that profitability targets are undervalued. Therefore, a negative adjustment occurs if no share of a private investor is retained.

The financial score is calculated by combining quantitative and qualitative aspects. The proportion of debt and the cost of debt determine the initial quantitative score. Due to the fact that in the case of Tsumkwe no debt capital was used to finance the mini-grid, the scoring starts with the value of 1 (**Table 9**).

The initial score can be changed through two adjustments, which are qualitative aspects in terms of independence and efficient management based on the funding. Major emphasis is therefore placed on grant funding and private sector participation.

		Share of debt			
		30%	50%	70%	100%
Cost of debt/revenues	**0%**	1	1	1	2
	10%	1	1	2	3
	20%	1	2	3	4
	40%	2	3	4	5
	60%	3	4	5	5
	80%	4	5	5	6
	100%	5	6	6	6

Table 9.
Finance initial score: assignment.

Experience shows that a grant reduces the willingness of management to use financial resources efficiently - as demonstrated by the failure of socialist systems in the past. As a result, funding by a grant leads to an increase of the risk score by +1.

On the other hand, private sector participation enhances independence from political decisions and encourages efficient management of resources. A private investor can be regarded as vital to avoid underestimation of the profitability targets. Therefore, the risk score increases by +1 if no private investor is involved in the mini-grid project.

In the case of Tsumkwe, there is neither a grant, as explained in the introduction, nor a private investor with commercial objectives. This leads to an increase of the risk score by +2 (see **Table 10**), resulting in a total financial risk score of 2.0, which indicates an overall moderate level of risk.

		Adjustment
Grant?	Yes	1
Private Investor?	No	1
Total adjustments		2
Final Financial Score		3

Table 10.
Finance adjustment.

4.5 Educational assessment

> This criterion is perceived as an assessment of management's ability and willingness to coordinate, communicate, and implement educational goals in order to maintain continuous economic performance and mitigate technical failures. The assessment considers only quantitative data, i.e., no qualitative adjustments are made. Ultimately, it is imperative that the management of a mini-grid is able to actively manage and contribute to technical quality through a planned educational policy. The quality of educational efforts is evaluated by the number of educational activities and the cost of educational activities compared to the total cost on average over the last three years.

When new technologies are adopted, as in the case of Tsumkwe, the shift from a diesel-only generator to a solar PV-diesel hybrid mini-grid system demands a significant amount of training for various groups of people to achieve better technical proficiency due to emerging productive use cases among the community [23, 30]. In fact, in order to operate an economic viable mini-grid system, productive use cases and the commercial usage of electricity produced are urgently needed, thus the engagement of the local community is a key aspect that needs to be considered already in the planning and development phase. In addition, a better understanding of electricity generation and use can potentially encourage people to replace inefficient equipment, both helping to save electricity costs in the long run and stabilizing the system. A series of training/education activities should therefore accompany every mini-grid project by default. In the case of Tsumkwe, the community was indeed involved and employed in the construction of the mini-grid. In addition, educational campaigns have been organized, focusing particularly on informing the community about the project itself, maintenance of stoves or solar water heater, as well as energy efficiency. This was done by distributing several flyers that provided information on how to save money through energy efficient measures, among other things [31, 32]. However, it now became apparent that the approach was not successful or rather not sufficient. Based on the results of a research trip to Tsumkwe in 2020, relatively few community members were aware that their prepaid electricity payments contained charges for service, maintenance, as well as a levy for government agencies. The remaining respondents were convinced that they were only paying for electricity, or indicated that they did not know what they were in fact paying for [23]. A field trip to a neighboring village, which was already conducted in 2016, presented similar results. A considerable number of people was unaware of solar energy and what it entails. Solely business people and those who already installed small solar panels on their rooftop were sufficiently informed [33]. This shows that the design of an educational component within the management of mini-grid systems is essential for the sustainability and the durability of the mini-grid system. As the case of Tsumkwe illustrated, this is not done by distributing informative material, but rather through direct contact with the community members, joint discussion about tariff considerations and training offerings.

From a technical point of view, regular preventive maintenance, which can only be accomplished with a trained team, ensures the optimized operation of a mini-grid and thus, enhances its technical sustainability. Apart from the educational requirements for community members, education and training of operating personnel is consequently essential to ensure the (economic) lifespan of the equipment for better service. At best, the responsible personnel is recruited locally. When the mini-grid system in Tsumkwe was expanded in 2011, two of twelve battery banks were obviously not properly connected to the other batteries. As a result, some of the additional battery capacity was unavailable. The fact that this problem was not identified until a research trip in 2019 clearly demonstrates the lack of knowledge and experience in operating off-grid power systems in the analyzed community.

The level of education proved to be a weak point within the five dimensions examined in the case of Tsumkwe. Based on our impression by a fact finding mission trip in July 2019 only one education measure per year was reported upon. Related to revenues expenses for education can be seen around 5% (**Table 11**). Due to the fact that in the municipality only very few educational measures were carried out that were useful in any way, combined with today's very low level of awareness regarding electricity among the population and poorly maintained facilities, this results in a value of 5 (**Table 12**), indicting a high risk level.

Component	Estimation
Number of educational measures p.a.	1
Cost of education	**5,0%**

Table 11.
Education score components.

		Number of educational measures				
Expenses education/revenues		20	15	10	5	0
	0%	1	2	3	4	5
	5%	2	3	4	5	6
	10%	3	4	5	6	6
	15%	4	5	6	6	6

Table 12.
Education: Scoring grid.

4.6 The overall rating

As alluded at the beginning, the five key components of a mini-grid rating which have been distilled into two profile categories – the "Regulatory framework effectiveness and economic profile" and the "flexibility and performance profile" – will be summarized and combined to produce a final, overall mini-grid rating. The two aforementioned profiles that were created as an intermediate step are the base to attain the final, overall SRMP mini-grid rating.

The five factors of the SRMP mini-grid rating procedure are allocated to two categories – the 'regulatory framework effectiveness and economic profile' and the 'flexibility and performance profile'. These profiles are subsequently combined to receive a final, overall mini-grid rating. **Table 13** presents the results in the case of the Tsumkwe mini-grid.

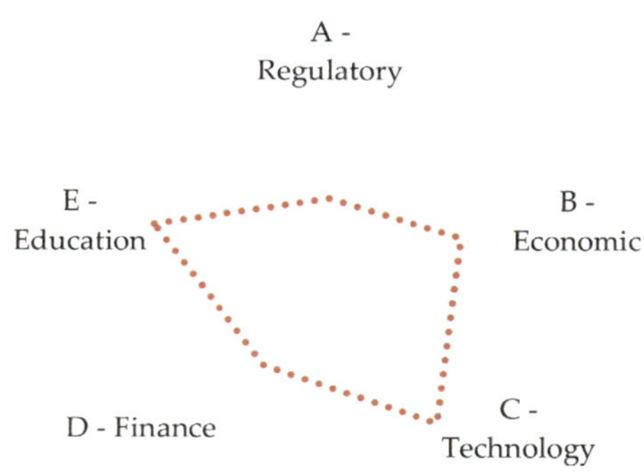

Factor	Score
A - Regulatory Framework	2.25
B – Economic	3.75
C – Technology	5.0
D – Finance	3.0
E – Education	5.0
Profiles	**Score**
Regulatory framework effectiveness and economic profile (A&B)	3.0
Flexibility and performance profile (C&D&E)	4.33
Total Rating	**bb+**

Table 13.
Final risk rating for the mini-grid in Tsumkwe.

		Regulatory framework effectiveness and economic score (A&B)										
Flexibility and performance profile (C&D&E)	**Category**	**Superior**	**Extremely strong**	**Very strong**	**Strong**	**Moderately strong**	**Intermediate**	**Moderately weak**	**Weak**	**Very weak**	**Extremely weak**	**Poor**
Category	**Score**	**1,0**	**1,5**	**2,0**	**2,5**	**3,0**	**3,5**	**4,0**	**4,5**	**5,0**	**5,5**	**6,0**
Extremely strong	1	aaa	aaa	aaa	aa+	Aa	a+	a	a-	bbb+	N/A	N/A
Very strong	1,8	aaa	aaa	aa+	Aa	aa-	a	a-	bbb+	bbb	bb+	bb-
Strong	2,3	aaa	aa+	aa	aa-	A	a-	bbb+	bbb	bb+	bb	b+
Moderately strong	2,8	aa+	aa	aa-	a+	a-	bbb	bbb-	bb+	bb	bb-	b+
Intermediate	3,3	aa	aa-	a+	A	bbb+	bbb-	bb+	bb	bb-	b+	b
Moderately weak	3,8	aa-	a+	a	bbb+	Bbb	bb+	bb	bb-	b+	b	b
Weak	4,3	a	a-	bbb+	Bbb	**bb+**	bb	bb-	b+	b	b-	b-
Very weak	4,8	N/A	bbb	bbb-	bb+	bb	bb-	b+	b	b	b-	b-
Extremely weak	5,3	N/A	bb+	bb	bb-	b+	b	b	b-	b-	ccc/cc	ccc/cc

Table 14.
Final Investment grade.

Taking all of the above dimensions into account, the risk rating results in a "moderately weak rating" (bb+). While the overall result reflects the general tendency towards a rather challenging environment, the final evaluation also revealed strengths and weaknesses.

Table 14 clearly illustrates that Tsumkwe's mini-grid is exposed to high risk associated with a low level of technology maintenance and a widespread lack of education.

The economic environment is assessed by a moderate risk (economic score B). That is a typical outcome for remote areas with limited access to income generating activities. The (national) regulatory environment is admittedly transparent and relatively stable. A shortfall is outlined in the ERI 2019, as Namibian regulations do not necessarily have a convincing impact on consumers, but rather on power utilities [25]. The conducted analyses, moreover, uncovered the substantial need to translate regulatory knowledge, technical guidelines and issued codes from large energy projects to much smaller mini-grid initiatives. While many policies target grid-based power generation, off-grid regulations are indeed scarce. The Africa Minigrid Developers Association, moreover, ascertained that the digitization of relevant processes would potentially accelerate the development [34]. Apart from the digitalization of e.g. licensing processes to simplify the establishment of mini-grids for investors, this also corresponds to the above-mentioned finding that online portals are needed to disseminate information concerning off-grid possibilities among the population. As the mini-grid in Tsumkwe received funding from donors, the financial score is rated to be less strong. Private investors have inevitably a positive impact on the profitability due to allegedly commercial targets.

5. Conclusion

This paper has extended the rating model based on the standardized risk management procedure (SRMP) developed by Steurer et al. 2017 (for more information see [6]). The SRMP approach is applied for the mini-grid in Tsumkwe, Namibia. The findings result in an assessment of the mini-grid in Tsumkwe with a moderate level of risk. Important for the strategic management of the mini-grid are the identified strengths and weaknesses supporting long-term risk management. In this case, there main potential is given in the fields of technology and education. Reducing the share of diesel would lead to improved earnings. Further, a well-organized replacement process is needed to ensure long-term commercial profitability. Particularly, there is evidence to expose the true costs for off-grid electricity distribution as well as to issue a central guideline for particularly off-grid tariff setting. Developing productive use cases to boost and restore the local economy is, furthermore, crucial, which, however, needs to be complemented with comprehensive training offers targeting both operating personnel and the population.

Acknowledgements

The authors acknowledge financial support received from the German Federal Ministry of Education and Research (BMBF) via the PROCEED project. Furthermore, we would like to acknowledge the great collaboration in the PROCEED project, especially with Özge Dolunay and Sven Kühnel, who have contributed to the presented findings to a large extent.

Conflict of interest

Both authors declare that they have no conflict of interest.

Author details

Inken Hoeck* and Elmar Steurer
Department Business and Economics, Neu-Ulm University of Applied Sciences, Neu-Ulm, Germany

*Address all correspondence to: inken.hoeck@hnu.de

References

[1] United Nations. The Sustainable Development Goals Report 2020 [Internet]. 2020. Available from: https://unstats.un.org/sdgs/report/2020/The-Sustainable-Development-Goals-Report-2020.pdf [Accessed: 2020-11-02]

[2] Global Commission to End Energy Poverty. 2020 Report: Electricity Access [Internet]. 2020. Available from: https://www.rockefellerfoundation.org/wp-content/uploads/2020/12/GCEEP-Electricity-Access-2020-Report.pdf [Accessed: 2020-08-05]

[3] IEA, IRENA, UNSD, WB, & WHO. Tracking SDG7: The Energy Progress Report 2019 [Internet]. 2019. Available from: https://irena.org/publications/2019/May/Tracking-SDG7-The-Energy-Progress-Report-2019 [Accessed: 2020-09-12]

[4] IEA. SDG7: Data and Projections [Internet]. 2020. Available from: h ttps://www.iea.org/reports/sdg7-data-a nd-projections [Accessed: 2020-09-12]

[5] IRENA, IEA, & REN21. Renewable Energy Policies in a Time of Transition [Internet]. 2018. Available from: https://www.irena.org/publications/2018/Apr/Renewable-energy-policies-in-a-time-of-transition [Accessed: 2020-09-14]

[6] Steurer E, Wagemann B, Manetsgruber D. Standardized Risk Management Procedure (SRMP) for Mini Grids Case Study: Mini-grid Sunderbans, West Bengal, India. 2019. Paper presented at the International Conference on Technologies and Materials for Renewable Energy, Environment and Sustainability, Beirut

[7] Hauser J. Namibia. Erneuerbare Energien zur dezentralen Energieversorgung. Zielmarktanalyse 2019 mit Profilen der Marktakteure. 2018. Houghton: Southern African – German Chamber of Commerce and Industry

[8] United Nations. UN Namibia goes solar, halving power consumption [Internet]. 2017. Available from: https://www.greeningtheblue.org/news/un-namibia-goes-solar-halving-power-consumption [Accessed: 2020-10-12]

[9] Electricity Control Board. ECB Publication on Electricity Supply Industry Statistical Bulletin Electricity Access in Namibia 2016 [Internet]. 2017. Available from: https://www.ecb.org.na/images/docs/Statistical_Bulletin/ESI_Stats_Bulletin_2016.pdf [Accessed: 2020-07-02]

[10] GTAI, GIZ, & AHKs. Partnership Ready Namibia: Renewable Energy. 2020. Eschborn: Deutsche Gesellschaft für Internationale Zusammenarbeit

[11] Government of Namibia. Vision 2030 Overview [Internet]. Undated. Available from: https://gov.na/vision-2030 [Accessed: 2020-07-02]

[12] Ministry of Mines and Energy. National Energy Policy [Internet]. 2017. Available from: http://www.mme.gov.na/files/publications/fd8_National%20Energy%20Policy%20-%20July%202017.pdf [Accessed: 2020-07-02]

[13] Solar Age Namibia. Tsumkwe Energy. Observation Trip Report [Internet]. 2005. Retrieved from Tsumkwe Energy: http://tsumkweenerg y.org/project-reports/observation-re port/ [Accessed: 2021-01-09]

[14] United Nations Development Programme. Off-Grid Energisation Master Plan for Namibia [Internet]. 2013. Available from: https://www.na.undp.org/content/namibia/en/home/library/environment_energy/off-grid-energisation-master-plan-for-namibia.html [Accessed: 2020-07-02]

[15] Zongwe DP, Ileka H, Reuther K. Rural Electrification with Hybrid Mini-grids: Finding an Efficient and Durable Ownership Model. 2017. The Law Reform and Development Commission of Namibia at 25: A Quarter Century of Social Carpentry (pp. 351-382).

[16] Klintenberg P, Azimoh CL, Wallin F. Successful technology transfer: What does it take? 2014. Applied Energy. doi:http://dx.doi.org/10.1016/j.apenergy.2014.01.087

[17] Ashton B, Bisacky L,Boyd D, López J. Lights on the Horizon: A Socioeconomic Impact Evaluation of Rural Electrification in Tsumkwe, Namibia [thesis]. Massachusetts: Worcester Polytechnic Institute; 2012.

[18] The International Bank for Reconstruction and Development. Mini Grids and the Arrival of the Main Grid: Lessons from Camodia, Sri Lanka and Indonesia [Internet]. 2018, Available from: https://openknowledge.worldbank.org/bitstream/handle/10986/29018/134326.pdf?sequence=6&isAllowed=y [Accessed: 2020-07-02]

[19] Ministry of Mines and Energy. National Renewable Energy Policy for the Republic of Namibia [Internet]. 2016. Available from: http://the-eis.com/elibrary/sites/default/files/downloads/literature/National%20Renewable%20Energy%20Policy%20for%20the%20Republic%20of%20Namibia.pdf [Accessed: 2020-07-02]

[20] Franz M, Peterschmidt N, Rohrer M, Kondev B. Mini-grid Policy Toolkit. Policy and Business Framework for Successful Mini-grid Roll-outs [Internet]. 2014. Available from: http://euei-pdf.org/downloads/ [Accessed: 2020-10-04]

[21] Schultz R, Schumann C. Off-Grid Energisation Master Plan for Namibia [Internet]. 2007. Available from: http://www.undp.org/content/dam/namibia/docs/enrgy&enviro/undp_na_ergyenviro_OGEMPReport_2007.pdf [Accessed: 2020-07-02]

[22] Chiguvare Z, Stockmayer M, Martonova L, Wetzer W. Nationally Appropriate Mitigation Action: Rural Development in Namibia through electrification with renewable energies [Internet]. 2015. Available from: https://www.undp.org/content/undp/en/home/librarypage/environment-energy/mdg-carbon/NAMAs/nama-on-rural-development-in-namibia-through-electrification-wit.html [Accessed: 2020-09-12]

[23] Hoeck I, Steurer E, Dolunay Ö, Ileka H. Challenges for off-grid electrification in rural areas. Assessment of the situation in Namibia using the examples of Gam and Tsumkwe. 2021. Energy, Ecology and Environment (unpublished).

[24] Osawa B, Telep P. How do we license it? A guide to licensing a mini-grid energy service company in Kenya [Internet]. 2015. Available from: https://www.giz.de/de/downloads/GIZ2015-ProSolar-Licensing-Guidebook.pdf [Accessed: 2020-10-06]

[25] African Development Bank. Electricity Regulatory Index for Africa [Internet]. 2019. Available from: https://africa-energy-portal.org/sites/default/files/2019-12/ERI%202019.pdf [Accessed: 2020-12-06]

[26] Wagemann B, Steurer E, Manetsgruber D. Photovoltaic Mini-grids in Remote Rural Areas of Developing Countries: Research on Strategies to Improve Operational Sustainability. 2019. Paper presented at the International Conference on Economics, Management and Technology, Neu-Ulm.

[27] Namibia Statistics Agency. The Namibia Labour Force Survey 2018 Report [Internet]. 2019. Available from: https://d3rp5jatom3eyn.cloudfront.net/

cms/assets/documents/Labour_Force_Survey_final_-_2018.pdf [Accessed: 2020-12-06]

[28] Namibia Statistics Agency. Namibia Household Income and Expenditure Survey [Internet]. 2016. Available from: https://d3rp5jatom3eyn.cloudfront.net/cms/assets/documents/NHIES_2015-16.pdf [Accessed: 2020-12-06]

[29] IRENA. Innovation Outlook: Renewable Mini-grids [Internet]. 2016. Available from: https://irena.org/-/media/Files/IRENA/Agency/Publication/2016/IRENA_Innovation_Outlook_Minigrids_Summary_2016.pdf?la=en&hash=B9B2B4E5FFB4248A1C4BE074B481E63B87C9C702 [Accessed: 2020-08-26]

[30] Inglesi-Lotz R, Diez del Corral Morales L. The effect of education on a country's energy consumption: Evidence from developed and developing countries. 2017. Economic Research Southern Africa.

[31] RECP, EUEI, ARE, REN21. Mini-grid Policy Toolkit- Case Study [Internet]. Undated. Available from: http://minigridpolicytoolkit.euei-pdf.org/system/files_force/attachments/Mini-grid%20Policy%20Toolkit%20Case%20Study%20-%20Namibia60ab.pdf?download=1 [Accessed: 2020-12-06]

[32] Tsumkwe Energy Project. Energy efficiency tips for Tsumkwe [Internet]. Undated. Available from: http://www.tsumkweenergy.org/project%20reports/flyer%203/ [Accessed: 2021-01-26]

[33] Haingura I H, David N P, Desney A, Nangolo E M. Assessing energy utility in the GAM community to implement the Off-Grid Energisation Master Plan. 2016. Namibian Energy Insitute (NEI) (unpublished)

[34] Economic Consulting Associates, Africa Minigrid Developers Association. Benchmarking Africa's Minigrids [Internet]. 2020. Available from: https://africa-energy-portal.org/sites/default/files/2020-08/AMDA%20Benchmarking%202020.pdf [Accessed: 2021-01-26]

Chapter 3

Smart Metering in Infrastructure-Less Communication Environments and Applicability of LoRa Technology

Biswajit Paul and Rajesh Palit

Abstract

Advanced-Metering-Infrastructure (AMI) is an integral part of Smart-Grids (SGs). It enables accurate consumer billing in presence of dynamic pricing, and improves efficiency and reliability of electricity distribution in presence of distributed generation. Value-added features of AMI such as diagnostics and maintenance service can identify the anomalous power consumption patterns of appliances at the end of their life cycle. Water and gas utility distribution networks in smart cities will incorporate AMI as an application of Internet-of-Things (IoT). The communication infrastructure plays a crucial role in enabling two-way communication between Smart-Meters (SMs) and the utility. AMI's bi-directional communication facility supports precise modeling of load information and data management system facilitating demand-response applications to reduce energy wastage. Researchers have investigated the role of wireless technologies in Home-Area-Networks (HANs), Neighborhood-Area-Networks (NANs) and Wide-Area-Networks (WANs) in AMI. The arrival of new Low-Power-Wide-Area-Networks (LPWANs) technologies has opened up new technology integration possibilities in AMI. However, it is essential to understand the AMI architecture, envisioned application types, network requirements, features and limitations of existing technologies to determine a technology's integration suitability in an application for smart metering technology. This chapter discusses LoRa for smart metering in infrastructure-less environments and the possible use of our multi-hop routing scheme.

Keywords: Smart-Grids, Advanced-Metering-Infrastructure, Smart-Meters, IoT, LPWANs, LoRa, Routing

1. Introduction

Advanced-Metering-Infrastructure (AMI) is an electronic system capable of providing more information than a conventional meter besides measuring energy consumption and supporting electronic communication to transmit and receive data [1]. Initially, SMs could measure the electricity used and generated, and remotely control the supply and cut-off whenever necessary. Automated monthly reads, one-way outage, tamper detection and uncomplicated load profiling were possible using one-way communication. AMI integrated two-way communication technology and upgrading SMs, including service switching, time-based rates, remote programming, power quality measure, user interface, etc. Current SMs should be capable of: (i) capturing electricity usage in a real-time and possibly distributed generation; (ii) providing local and remote meter reading; (iii) remotely controlling the meter and even cut off the supply and (iv) providing facilities to link with other commodity supply such as gas and water [2]. AMI's two-way communication system will enable some services such as collecting data related to the power grid status and the delivery of commands in the electricity distribution grid [3]. A wide array of energy resources, including renewable energy producers and mobile energy storage, can be linked and utilized by SGs' infrastructure [4]. Since non-renewable energy resources are limited, energy-efficient and effective use of renewable resources to make smart energy become crucial for social and economic developments, and the SGs are considered a key enabler for smart energy [2]. Acquiring accurate information, automated decision support and handling events in a timely fashion are required for delivering smart energy, which is not possible without SGs.

The SGs can be used for (i) improving situational awareness and operator assistance; (ii) enhancing reliability using autonomous control actions; (iii) enhancing efficiency by maximizing asset utilization; (iv) designing superior defense mechanism against malicious attacks; (iv) incorporation of renewable resources into power grids; (v) integration of different types of energy storage and other resources; (vi) bi-directional communication between the consumer and utility; (vii) enhancing market efficiency and ensuring a higher quality of service to influence an increasingly digital economy [5]. The utility companies can detect peak load demand and control them via the Demand-Response (DR) program using the load management system [6]. In the DR program, SMs enable customers to control energy use within the HANs, and act as active participants in overall grid load management [6]. The outage notification using the SG metering network helps the whole grid system efficiently respond to the power outage condition of SMs [6]. It is possible to identify the equipment, communications and processes for Electric-Vehicle's (EV) charging control [7]. Information-and-Communication-Technologies (ICT) planning approach can be combined with distributed network planning processes and tools. New business models for SGs' can be generated to support customer-side participation, increase penetration of renewable energy sources, and foster the electricity market's participation [7].

Smart metering is one of the most promising applications of IoT, foreseen by some industries and researchers. In AMI, IoT can be utilized to collect data, measure abnormality in the SGs, exchange information among SMs, monitor electricity quality and distributed energy, and analyze user consumption patterns. Interactions among users and SGs, enhancing SG services, meeting marketing demand, improving Quality-of-Service (QoS), controlling smart appliances and monitoring renewable energy sources can be made feasible in a smart home. The monitoring center in EV assistant management systems will manage car batteries, charging equipment, charging stations and optimize resources. SGs' abilities such as processing, warning, self-healing, disaster recovery, and reliability can be enhanced using the comprehensive sensing and processing abilities of IoT [8]. The extensive scale deployments of network-connected smart electricity meters in households and commercial/industrial locations represent IoT devices [9]. The forecast on a large-scale IoT network deployment indicates potential challenges such as connectivity and traffic requirements for resource-constrained wireless networks. The heterogeneous nature of traffic, e.g., static, intermittent, delay-sensitive, delay-tolerant, small or large packets, will make wireless network design more complicated. The integration of various IoT enabling technologies will be necessary to address the connectivity issue in Machine-Type-Communication (MTC). According to Ericsson, there will be 28 billion smart devices by 2021, and 15 billion of them will be connected through Machine-to-Machine (M2M) and consumer electronic devices [10]. 7 billion of those devices will be connected through cellular and LPWAN technologies.

It is essential to guarantee communication between SMs and utility companies for AMI to become an efficient and reliable SG component. Besides technical and economic considerations, issues such as customers' geographic location, population density, and reducing connectivity costs using existing communication infrastructure, especially in the last mile, will have to be considered in providing full coverage to the SMs [11]. Integration of millions of stakeholders in rural/remote areas globally requires developing last-mile telecommunication technologies. Providing cost-effective last-mile connectivity to rural areas remains a challenging issue as rural/remote areas are characteristically influenced by factors such as scattered user base, resistance to adopt new technology, and affordability. Also, economic and social disparities across regions lead to a digital divide between urban and rural/remote locations resulting in a lack of access to computing infrastructure. Several factors are crucial in determining a viable technology solution for last-mile connectivity in rural scenarios, such as geographic location, economic condition, motivation/incentive and adaptability, sustainable business framework etc. [12]. However, QoS requirement analysis of service is important in determining viable technologies in infrastructure-less environments. As an example of varying QoS requirements, two categories of services are required in Power-Wireless-Private-Network (PWPN): (i) latency tolerant services whose characteristics are low data rate and massive access devices, and (ii) the mission-critical services which require the data transmission in real-time or near real-time, such as distribution automation [13].

Due to low installation costs, the possibility of rapid deployment over wide areas, improved data rates and network capacities, increasing portions of utilities' AMI systems will rely on wireless communication technologies [3]. There is no single communication technology that can meet the requirements of all the diverse AMI deployment scenarios [3]. Narrowband transmission is more suitable than wideband transmission on the consideration of hardware cost for some SM applications that require infrequent packet transmissions [13]. In [13], almost similar performances were achieved for two popular narrowband transmission technologies, NB-IoT and LoRa but according to the authors, NB-IoT may be more suitable for the narrowband PWPN considering the cost, evolution and possibility of integration into the LTE system. According to LoRa Alliance, LoRaWAN can complement LTE variations in serving different application segments to achieve a + 20 dB gain over the legacy cellular system in providing extended network coverage [14]. LoRaWAN has been explored in the context of an SM in [15–17]. In [15], the authors found that LoRaWAN was scalable, configurable, the setup process was easy, and performed well for real-time smart metering applications. An SM prototype was developed in [16] to monitor the energy efficiency of induction motors. Enhancing the security of meter-reading systems based on LoRa technology through an improved session key management was proposed in [17].

Multi-hop wireless technologies' capabilities for routing traffic from the SMs to the Universal-Data-Aggregation-Points (UDAPs) will play an important role in ensuring full network coverage [11]. Although explored for harsh battlefield environments [18], where access to the environment may be limited with no preexisting communication infrastructure; employing relay nodes in acquiring a larger coverage area for applications with relaxed QoS requirements can be explored for similar SM application scenarios. The required number of SMs should be large enough to provide multi-hop connectivity to ensure 100% coverage when UDAP capacity is low [11]. Also, for small populations and high UDAP capacity, full connectivity to the SMs cannot be ensured unless the number of SMs is large enough to enable multi-hop connectivity [11]. Designing a routing protocol for AMI is not easy as typical SMs are resource-constrained embedded devices with limited processing power and storage capabilities. The links between the devices are generally characterized by high packet loss rates, low bandwidth, and instability due to unplanned network deployments and the use of low-power link-layer technologies [3]. Discussion on a scalable route map for AMI to achieve target coverage of SMs with a reduced cost in terms of technological resources can be found in [11], where the authors emphasized optimal routing in heterogeneous wireless networks.

SMs are integral components of SGs and future smart cities. However, there are diverse sets of requirements for various potential applications of SMs. It is important to understand different application types, varying requirements and enabling technologies for determining suitable technologies under different circumstances. LoRa technology, according to our literature review, can be considered for potential last-mile SM applications with relaxed QoS requirements. While low-cost deployment and extended network coverage are the most attractive features of LoRa technology, careful examination of SM application requirements, and coverage and transmission reliability of LoRaWAN are required. Multi-hop routing can improve certain issues in LoRaWAN such as improving energy efficiency and network coverage. Although the performance of our multi-hop routing scheme was explored for LoRa technology, its possible use in other wireless mesh and cognitive networks in improving certain parameters requires further investigation. The chapter is organized as follows: AMI architecture, applications, requirements and enabling technologies are explored in section-II. Technological features of LoRaWAN and our developed multi-hop routing scheme [19] are presented in section-II. Conclusions are drawn in Section-IV.

2. AMI and enabling technologies

The Pioneering works of Smart-Grid-Architecture-Model (SGAM) focused on (i) creating a methodology to elicit applications for future and emerging SGs and (ii) reaching a solution and a blueprint for technical architectures of future technology portfolios. The massively interconnected, complicated SG system is a complex system with the diversity, variety and scale of its constituent subsystems and a specific dynamic and alterability of its structure [7]. The SGAM acts as a reference designation system, and the three principal axes of the dimensions are: (i) value creation chain ("Domains"), (ii) automation pyramid ("Zones") and (iii) interoperability [7]. The individual domains in the energy conversion chain can be described as: bulk generation, transmission, distribution, distributed energy resource and customer premises. The individual zones representing ICT to control the energy conversion chain can be described as the market, enterprise, operation, station, field and process. The interoperability layer includes the business layer, function layer, information layer, communication layer and component layer. SGAM allows studying the system across several domains such as electrical power systems and communication infrastructures, and in this chapter, our focus is on communication infrastructures.

2.1 AMI architecture

AMI components are a central system, two-way communication networks, data concentrators, and SMs [8]. Generally, communication in AMI can be divided into

[20]: (i) Wide-Area-Networks (WANs), (ii) Neighborhood-Area-Networks (NANs) or Meter-Local-Area-Networks (MLAN) [6] and (iii) Home-Area-Networks (HANs). The HANs provide connections among distributed energy resources, Gateways (GWs), EVs, SMs, etc., while the NANs facilitate connections for several SMs that need to send their data to the corresponding data concentrator. SMs are installed at customers' locations or other positions in SGs, which need to send data to the central system. A group of appliances, entertaining systems, lighting systems, energy storage and generation (solar, wind etc.), electric vehicles constitutes HANs where a smart meter acts as a home GW that links the HANs with the NANs [4] or in other words, transmit collected data to the utility and receive control commands from the utility [6]. The NANs contain several homes supplied by one transformer, and thus, they should carry a massive volume of data with satisfactory service requirements [8]. Data from SMs are aggregated and compressed in uplink connection, and data are relayed to SMs in downlink connection by the data concentrator. Despite increasing SMs' data transmission time, data concentrators improve the scalability and reliability of the SGs, reduce SMs' power consumption and data collision. In [6], The MLAN refers to the communication between the SMs and the data concentrator and is mainly located at (i) the energy distribution domain comprised of the data concentrators and (ii) the field area network that includes devices such as monitors, re-closers, switches, capacitors, controllers etc. The WANs provide connections among some data concentrators and the central system. SMs' data are collected, stored, analyzed and processed by the central system, which may have several components such as Meter-Data--Management-System (MDMS) [8, 6], geographic information system, outage management system, consumer information system, power quality management, and load forecasting system [8]. The communication architecture in AMI is shown in **Figure 1**.

Several proposed IoT architectures for integration into the SGs were discussed in [8]. In a three-layers-architecture, Layer-1 includes SMs, network devices, and communication protocols. Devices that are responsible for data reception at the central system constitute Layer-2, while artificial intelligent systems that provide information to decision and billing systems make up Layer-3. Some researchers recognized the three layers as the perception layer, network layer and application layer. Different types of sensors, such as GPS devices or cameras, are used in the perception/device layer. Different types of wired and wireless communication networks and the internet constitute the network layer responsible for mapping sensors' information to communication protocols [8] and delivering the mapped data to the application layer. Information received from the network layer is processed

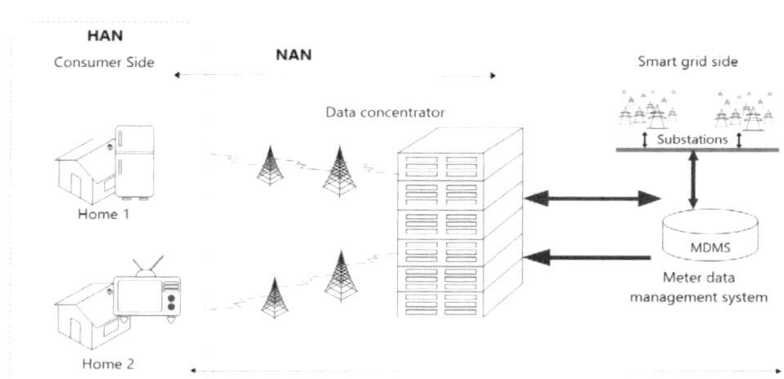

Figure 1.
AMI architecture.

by the application layer to monitor IoT devices. Some researchers proposed a four-layers-structure that included a cloud management layer and the device layer, network layer and application layer. Data storage, analysis, data and user management, etc., are performed in the cloud management layer. The device layer was further subdivided into two layers: the thing layer, which contained sensors, SMs, actuators, etc. and the gateway layer, which contained microcontrollers, communication modules, storage, etc. [8]. In another work, a four-layer model was divided into a terminal layer, filed network layer, remote communication layer, and master station system layer. Remote terminal units, SMs, etc., made up the terminal layer while different wired and wireless technologies constituted the field network layer. The control systems for generation, transmission and distribution in SGs were included in the master station system layer.

2.2 AMI requirements

Although smart metering technology opens up new possibilities, there are some specific issues, e.g., security, privacy, scalability, energy-efficiency, that need to be addressed.

Addressing the SGs' cyber-security aspect is extremely important, specifically the communication mechanisms that deal with the distribution subpart [4]. SG and smart home security were presented in [21], where the authors discussed some representative threats and evaluated theoretical impacts from smart homes to SGs and vice versa. The authors in [22] discussed data generation security, data acquisition security, data storage security, data processing security and security analytics in the SG networks and analyzed the suitability of various security analytics techniques, [23]. IoT-enabled cyber-attacks in several critical infrastructures, including SGs, were discussed in [24].

The energy consumption data collected by SMs can reveal sensitive consumer information, and thus, privacy is a key concern and a significant inhibitor of real-time data collection in practice [25]. The SMs' data can be used to invade consumers' privacy through disclosing information such as household occupancy or economic status. The authors in [6] argued that consumers' privacy could be breached at the network level as the SMs' data travels through insecure networks. Privacy was defined by information leakage rate in [26], and the impacts were explored in the context of renewable energy sources and energy storage devices. In [27], the authors quantitatively measured the information leakage of appliances' status from any reading stream and proposed a privacy-preserving streaming algorithm. Breach of privacy is a concern in IoT networks too [28].

A system's ability to gracefully handle growing amounts of work depends on effectively addressing scalability, which can be measured in various dimensions such as administrative scalability, functional scalability, geographic scalability, and load scalability [29]. AMI in SGs is an example case of a cyber-physical system, and an AMI communication infrastructure has to collect and process a massive amount of data from hundreds of thousands of SMs [29, 30]. Therefore, scalability is one of the most significant considerations for the AMI deployment in the SGs. Many deployed AMI systems collect data from SMs every 15 minutes [29]. However, more frequent communications are expected for advanced distribution automation and advanced asset management.

Communication reliability is one of AMI's fundamental blocks, particularly in the NANs, which transport traffic of varying requirements [31]. In some cases, the NANs' traffic sources such as SMs are required to collect time-based data for real-time, reliability-sensitive applications from customers. Interference and channel errors associated with IEEE 802.11 and the nature of SGs make the static multi-hop wireless mesh network a potential technology for AMI applications [31]. If a wireless mesh network is used, it must guarantee the QoS level required by respective application traffic to deliver an acceptable level of reliability and latency [31]. Determining a suitable data collection strategy and GW location in a wireless mesh network is essential to meet some SG performance requirements [32].

Green wireless communication which is essential to reduce environmental impacts of CO2 emissions from increased energy consumptions requires energy-efficient smart metering techniques in SGs [33]. ICT is responsible for 3% of energy consumption, and 2% of CO2 emission worldwide [34] and the AMI networks' energy consumption may worsen the issue [34].

2.3 AMI enabling technologies

The information flow in AMI requires communication technologies to establish and maintain the information exchange among different components. Each technology comes with its features and limitations, and integrating technology in AMI requires careful evaluation of application requirements. Organizations such as the NIST and the IEEE have been developing interoperability standards in SGs as the grid has to incorporate electrical, electronic, power electronic and networking components [35]. Three major infrastructure categories are observed in a typical telecommunication network: access network, metropolitan area network and core network [12]. The access network connects end users to the rest of the network, and it typically spans a few kilometers while the backhaul network consisting of metropolitan and core networks, connects the access network to the rest of the network. The metropolitan network covers of a few tens to a few hundreds of kilometers and the core network provides global connectivity. In the access segment, Power line communication (PLC) can be a viable solution for last-mile connectivity as it provides high-speed transmission facilities in houses or offices, and reduces the telecommunication capital expenditure [12]. Copper or fiber based wired telecommunication technologies provide higher data rates, and unlike wireless technologies, they are less susceptible to interference, signal loss, and Line-of-Sight (LOS) requirements. However, most deployments to provide last-mile broadband access in rural areas are based on wireless technologies due to their cost effectiveness, flexibility, and ease of installation, especially in challenging terrains while the backhaul connection from the wireless terminal to the core network uses wired connectivity [12].

Some technologies and their limitations for achieving last-mile connectivity have been discussed in [12]. WiFi is a mature technology due to commercialization but the IEEE 802.11 Medium-Access-Control (MAC) protocol give poor end-to-end performance for long range communication. Cellular networks can be an efficient last-mile solution for rural areas due to significant cellular penetration in many rural areas across the world and ability to provide voice as well as data connectivity. Small low-power cellular base station, called a femtocell, can be used to provide cost-effective cellular connectivity within its coverage range. LTE, a 4G cellular technology, allows high user mobility and can extend the battery life of mobile terminals. WiMAX supports broadband applications in both LOS and non-LOS (NLOS) as well as provides large coverage, and deployment of a WiMAX network is much cheaper than deployment of an LTE network for last-mile connectivity in rural areas. Low-cost connectivity alternatives which employ store and-forward networking are suitable for rural areas with minimum or no existing infrastructure and time-insensitive applications where basic data communication is more important than a large response time. Cognitive radio technologies can achieve large coverage with non-LoS links in last-mile connectivity while resolving cost-demand mismatch in rural areas utilizing unused licensed spectrum.

Narrowband power line communications and wireless communications are two leading technologies for SG applications as the overall system reliability can be enhanced by exploiting the diversity achieved from the simultaneous transmission of the same signal over power line and wireless links [35]. The Low-cost-deployment and multiple functionalities of WSNs make them an attractive technology for AMI applications, and a typical application of WSNs is remote power line monitoring. However, ensuring QoS requirements of SM applications is a challenging task for WSNs. A ZigBee-based communication for multifunctional electronic-current-transformers used in overhead and underground line monitoring was discussed in [36]. The optimized configuration of the WiGrid profile, an amended form of WiMax, was studied in [37], where the authors discussed the choice of frame duration, type-of-service to traffic mapping, scheduling strategies and the system architecture to meet the SGs' communication requirements. In the context of SMs, discussions on cognitive radio networks and satellite communications can be found in [38, 39], respectively.

SMs are typically connected to the Distribution-System-Operators (DSO) backend system using either: 1) a concentrator gathers the data from the SMs in its neighborhood using Wi-Fi or PLC connections, and then relays it using cellular or a wired connection to the DSO backend; or 2) Each SM sends data to the DSO backend using cellular network [9]. Although the concentrator based approach reduces the load of access network by aggregating data locally, it is not suitable for real-time applications [9]. Some of the wireless technologies which can be used in HANs include IEEE 802.15.4 (e.g., ZigBee and Zwave), IEEE 802.11 (WiFi) [6] etc. PLC has been the primary choice for communication between the SMs and data concentrators. However, wireless mesh networks in AMI have been proposed and deployed widely [6]. IEEE 802.16 (i.e., WiMAX), IEEE 802.20 (MobileFi), PLC, IEEE 802.11 (WiFi) and IEEE 802.15.4 (ZigBee) are some of the potential technologies for providing communication in WANs [6]. LTE as a NAN technology can transfer data to the utility in two ways: (i) using existing mobile network architecture of established mobile network operators and (ii) utilizing specialized network core architecture [4].

Technology	Frequency Band	Range	Maximum Data Rate	Channel Bandwidth
LoRa	868 MHz, 915 MHz	15 km	30 kbps	125, 250, 500 KHz
SigFox	915 to 928 MHz	20 km+	100 bps	100 Hz
INGENU-RPMA	2.4 GHz	15 km	20 kbps	1 MHz
eMTC	700–900 MHz	< 15 km	1 Mbps	1.08 MHz (1.4 MHz carrier bandwidth)
NB-IoT	700–900 MHz	< 35 km	DL: 170 kbps UL: 250 kbps	180 kHz (200 kHz carrier bandwidth)
EC-GSM-IoT	800–900 MHz	< 15 km	74 kbps (GMSK), 240 kbps (8 PSK)	0.2 MHz
Bluetooth	2.4 GHz	50 m	2 Mbps	2 MHz
ZigBee	868 MHz, 915 MHz and 2.4 GHz	Typically less than 1 km	250 kbps	2 MHz
Wi-Fi	2.4 GHz, 5 GHz	100 m	54 Mbps	22 MHz

Table 1.
Features of some wireless technologies.

The traffic profile generated by SMs can be categorized as M2M traffic as it consists of transmissions of small amounts of data from a very high number of devices [9]. Uplink communication is more challenging than downlink communication for M2M traffic for the commonly perceived application scenarios. LPWANs are considered suitable for massive IoT applications such as logistics, utilities, smart cities, consumer electronics, smart buildings, environment, agriculture and industry. LoRa, Sigfox, Ingenu Random Phase Multiple Access (RPMA), DASH-7 are some of the non-cellular-based LPWAN technologies. IEEE working group 802.11ah enhanced communication development for Bluetooth Low Energy 4.0, ZigBee and WiFi/IEEE802.11 to support short-range communication for MTC [10]. On the other hand, EC-GSM-IoT, NB-IoT, LTE Cat-M1 are cellular-based LPWAN technologies. In terms of addressing key network performance indicators, cellular-based technologies having better resources can outperform the non-cellular-based LPWAN technologies in most cases. **Table 1** summarizes some of the critical features of some enabling IoT technologies [40].

3. LoRaWAN and multi-hop routing

The communication protocol and system architecture for the LoRa networks are covered by LoRaWAN. Researchers investigated LoRaWAN from several different perspectives. Theoretical analysis of LoRaWAN scalability for European and North American bands was presented in [41, 42]. A robust physical layer in LoRaWAN ensured better scalability than pure ALOHA as some packets could be recovered under concurrent transmissions as conditions were met [43]. They claimed that packet loss depends on transmissions' timings, the Received-Signal-Strength-Indicator (RSSI) of the interferer, and the interfered transmission. However, in contrast to the notion of immunity of a particular data rate transmission from interference produced transmissions with different data rates, the authors in [44] found transmissions with different data rates in the same frequency channel could negatively impact each other. Packet delivery rate for each sensor improved as the sensors used relay nodes in the absence of direct communication between End-Devices (EDs) and the GW. The feasibility of concurrent transmission in the LoRa multi-hop network was studied in [45]. The authors in [46, 47] studied localization in LoRaWAN. The authors in [48] focused on electricity grids, where LoRa nodes were used as SMs that sent the average power demanded by their respective households during a given period. The low-cost transceivers' capability to schedule the frame transmissions, and long-term clock stability of nodes and packet forwarders were explored in [49], and the authors found LoRa networks to be suitable for IoT applications such as smart metering, smart building, and process industry.

Although LoRa is a candidate technology for last-mile coverage in applications with relaxed QoS requirements, some issues such as reliability of packet transmissions and dependence of achievable transmission range on network environments and data rates in LoRaWAN deserve attention. The node distribution scenario in LoRaWAN could be imagined by placing concentric circles around the GW, where the maximum distance supported by an SF served as the radius of each circle [50]. The authors in [51] found that Packet-Error-Rate (PER), maximum transmission distance and payload sizes are not independent factors in LoRaWAN while conducting experiments with 250-kHz-LoRa-channels. 10-bytes-payload Packets were successfully delivered with zero PER up to 8 km, while 100-bytes-payload Packets were successfully transmitted up to 6 km with near-zero PER. In one reported coverage test in [51] resulted in 2.2 km connectivity in one direction and 1.6 km in another direction with a hill in between the end-devices (EDs) and the

Gateway (GW). Placing the GW at the top of a tall building (19 floors), another reported experiment in [51] could receive packets up-to 2 km. At present, single-hop routing (i.e., direct transmission from EDs to the GW) is used in LoRaWAN for EDs-to-GW communication. It is expected that a single GW in LPWAN networks will handle several hundreds of non-rechargeable battery-powered EDs. Arbitrary node distributions in LPWAN applications make the non-uniform nature of energy consumption unavoidable as some nodes can be a few hundred meters away from the GW, while other nodes could be a few kilometers away. A significant amount of energy is consumed in packet transmission and reception, and nodes with large transmission distances become the critical nodes in the LPWANs under single-hop routing. It was argued in [19] that critical nodes, which have higher energy consumption rates compared to other nodes, are critical for network lifetime. Researchers are looking for feasible multi-hop routing schemes for extended network coverage and improving network lifetime.

Multi-hop routings in LPWANs are required to avoid complex algorithms that need huge computational facilities in EDs, to facilitate low-cost network deployment. Flexible node distribution scenarios, network connectivity, network coverage were addressed by our Extended-DRESG routing scheme [19]. Better performance in terms of energy efficiency was achieved with a multi-hop routing model over single-hop routing while maintaining a high network connectivity level. Payload aggregation in Extended-DRESG offered increased energy efficiency. Necessary calculations and decisions for routing in Extended-DRESG can be performed in the network planning stage, which is a desired feature for low-cost-network-deployment.

3.1 LoRaWAN

LoRa has two distinct layers (i) a physical layer using the Chirp-Spread-Spectrum (CSS) radio modulation technique and (ii) a MAC layer protocol (LoRaWAN). However, LoRa is widely known as a physical layer technology that utilizes a proprietary spread spectrum technique to modulate the signals in the sub - GHz ISM band [52]. The bidirectional communication in LoRaWAN is provided by the CSS technique, which spreads a narrowband input signal over a wider channel bandwidth. Transmitters generate chirp signals that vary of various frequencies over time without changing their phase between adjacent symbols. The need for expensive components to generate a stable local clock in a LoRa node is eliminated due to more uncomplicated and more accurate timing and frequency synchronization by ensuring phase continuity between different chirp symbols preamble part of the physical layer packet [10]. A severely attenuated signal several dB below the noise floor can be decoded by a distant receiver if the frequency change is slow enough to put higher energy per chirp symbol. The payload can range from 2255 octets for each transmission, and packet transmission is similar to ALOHA. LoRaWAN provides the required medium access control mechanism for communication among many EDs and GWs. LoRa MAC design aimed to mimic the IEEE 802.15.4 MAC interface to accommodate some major protocols such as 6LoWPAN and Constrained-Application-Protocol (CoAP). A good overview of LoRaWAN can be found in [52].

Three types of devices are found in a LoRa network: (i) EDs, (ii) GW(s), and (iii) LoRa network server. A star-of-stars topology is formed as EDs communicate with the GWs, using LoRa modulation following LoRaWAN and GWs forward the EDs' frames to a network server decoding and regeneration over a backhaul interface typically through Ethernet or 3G. IEEE 64bit Extended-Unique-Identifier (EUI) is used to assign IPv6 addresses to LoRa nodes. LoRaWAN ensures security by employing several layers of encryption using (i) a unique network key for the network layer, (ii) a unique application for end-to-end security at the application layer and (iii) a device-specific key for a node to join the network. EDs can be activated either via Over-TheAir-Activation (OTAA) or via Activation-By-Personalization (ABP). EDs need to be personalized first with the required information: a globally unique enddevice identifier (DevEUI), the application identifier (AppEUI) and an AES128 key (AppKey) before the OTAA process allow Eds to join the network and exchange data with the network server. LoRaWAN supports ClassA, Class-B and ClassC devices to address various application requirements and bi-directional traffic is controlled in different manners in networks utilizing different device types [10]. GW works as a bridge between end EDs and network servers. More than one GW in a LoRa network and multiple GWs can receive a single packet.

SFs, coding rate, signal bandwidth, and Adaptive-Data-Rate (ADR) are the factors that impact LoRa link design. Spread-spectrum modulation is performed by representing each bit of payload information by multiple chips of information [52]. The rate at which the spread information is sent is represented by symbol rate (RS). The ratio between the nominal symbol and chip rates defines SFs, and different SFs can send different numbers of symbols per bit of information. LoRa supports seven SFs, while SF7 to SF12 is used for uplink transmission. Four different cyclic error coding rates are supported in LoRaWAN for forward-error-detection and correction. LoRa also supports different bandwidth options for channels such as 125-kHz-LoRa-channel, 250-kHz-LoRa-channel, 500-kHz-LoRa-channel, 250-kHz-GFSK-channel. Higher signal bandwidth ensures a higher effective data rate, which reduces transmission time at the expense of receiver sensitivity. The ADR scheme allows EDs to transmit on any available channel using any available data rate, which can maximize EDs' battery life and improve network capacity. Currently, there is no time restriction on EDs or GWs for packet transmission. However, there is duty cycle (European band) and dwell time (North American band) restrictions depending on geographical locations. The Time-on-Air (ToA) is the time interval between the first bit of a message frame leaving the ED and the last bit of that message frame leaving the ED.

There are a preamble, an optional header and the data payload in a LoRaWAN packet [52]. The preamble is responsible for synchronization between the receiver and the incoming data flow. The preamble length is extendable as required by applications, while the default configuration is a 12 symbol long sequence. The receiver, which periodically restarts, undertakes the preamble detection process, and the receiver's preamble length should be configured identically to the transmitter preamble length. The maximum preamble length should be programmed on the receiver side if the preamble length is a variable or cannot be known beforehand. There are explicit and implicit header modes available in the packet header. The default mode of operation is explicit header mode, where information about the payload length, the forward error correction code rate and the presence of an optional 16 bits CRC for the payload are provided. Implicit header mode can be chosen to reduce transmission time if the payload length, coding rate and CRC presence are fixed or known in advance. The packet payload length can vary, and the payload contains the actual data coded at the error rate.

3.2 Multi-hop routing

The Extended-DRESG Scheme considered single-gateway-LoRaWAN-deployment with the GW placed at the center of the circular network coverage, and the maximum achievable transmission distance of the SX1272 transceiver is considered the radius of the circle [19]. The optimal single GW placement for constant and variable data rates was studied in [53], and the GW placement at the center of the single-gateway-LoRaWAN-deployment appeared logical and consistent with other available works. The routing scheme in [19] was evaluated for uniform node distribution. In multi-hop routing, nodes can use a different number of hops to reach the GW instead of single-hop routing. An evaluation framework evaluates the best ring-hop-combinations from R! combinations following extensive simulations, where R clusters are formed through virtual rings. The critical nodes in the optimal-ring-hop combination consume less energy than the critical nodes in any other ring. Network connectivity is an important issue that is often ignored. However, the Extended-DRESG scheme ensured at least 99% network connectivity for simulation results. The work-flow of Extended-DRESG is presented in **Figure 2**.

The total energy consumptions for nodes in different rings for different routing processes, for example-case of four-rings, were presented in **Figure 3(a)**. Among Direct-Hop (DH), Next-Ring-Hop (NRH) and Variable-Hop (VH) routings, NRH performed the worst, while VH routing can almost double the network lifetime if used of DH routing instead. The impact of the number of rings on network lifetime for different routing processes was shown in **Figure 3(b)**. When rings were placed following the 'Equidistant' distance spreading model, VH routing achieved significant energy efficiency over DH routing, for example, 96% and 67% improvements for selecting 8 and 4 rings, respectively. However, 3 or 4 virtual ring selections appear more practical considering other network-related issues. If network connectivity was not given priority, VH routing could be even more energy-efficient, reflected in **Figure 3(c)**. Also, it was shown that payload aggregation could further improve network lifetime in VH routing in **Figure 3(d)**. For the example case of four rings, network lifetime in VH routing was improved by approximately 1.67 times with payload aggregation.

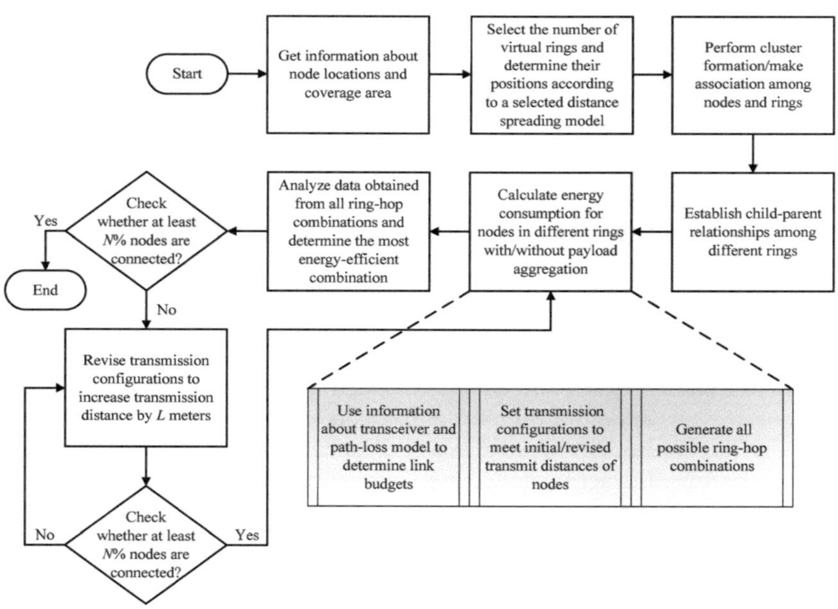

Figure 2.
Performance evaluation framework.

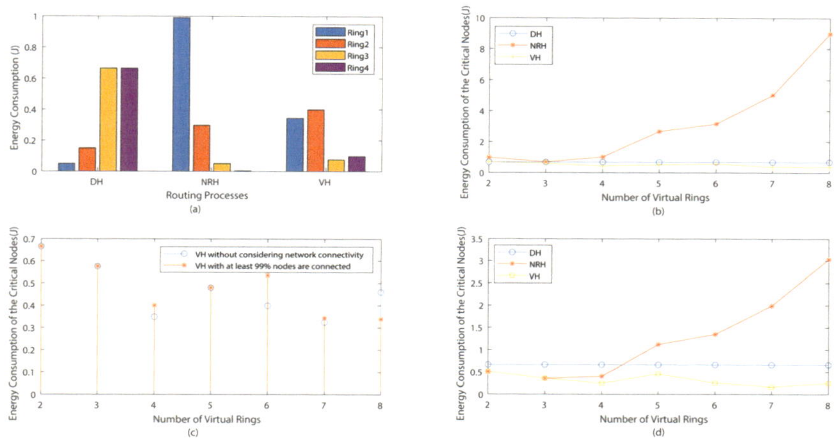

Figure 3.
Performance of extended-DRESG. (a) Energy consumption of critical nodes, (b) Impact of number of virtual rings, (c) Energy compensation to increase network connectivity and (d) Impact of payload aggregation.

4. Conclusions

This chapter discussed the communication infrastructure for AMI architecture and features of different wireless technologies in determining their suitability in HANs, NANs and WANs. Security, privacy, scalability, reliability, and energy efficiency of AMI are vital concerns in implementing smart metering technology. The suitability of wireless technologies in HANs, NANs or WANs depends on its ability to effectively address the concerning issues. Besides, the integration of smart metering technology in IoT demands exploring existing and new technologies carefully as lack of comprehensive analysis in the presence of an enormous amount of information available in literature makes the process complicated. However, technology choice may also depend on geographical locations, accessibility to existing wireless technologies, infrastructure availability, and many more. LoRa technology has been explored in detail in this chapter for its potential use in AMI. Based on our observations, LoRa can be a potential replacement where lack of appropriate infrastructure hinders using other more appropriate technologies. Several issues have been identified in the literature that will have to be addressed by LoRa technology to become more inclusive. The Extended-DRESG multi-hop routing scheme can help LoRa address some of these issues such as network lifetime, coverage and connectivity.

Acknowledgements

This project was supported in part by University Research Center, SUST, under the research grant: AS/2020/1/31.

Abbreviations

AMI	Advanced-Metering-Infrastructure
SGs	Smart-Grids
IoT	Internet-of-Things
SMs	Smart-Meters

HANs	Home-Area-Networks
NANs	Neighborhood-Area-Networks
WANs	Wide-Area-Networks
LPWANs	Low-Power-Wide-Area-Networks
DR	Demand-Response
EV	Electric-Vehicle
ICT	Information-and-Communication-Technologies
QoS	Quality-of-Service
MTC	Machine-Type-Communication
M2M	Machine-to-Machine
PWPN	Power-Wireless-Private-Network
UDAPs	Universal-Data-Aggregation-Points
SGAM	Smart-Grid-Architecture-Model
MLAN	Meter-Local-Area-Networks
GWs	Gateways
MDMS	Meter-Data-Management-System
WSNs	Wireless-Sensor-Networks
PLC	Power-Line-Communication
LOS	Line-Of-Sight
MAC	Medium-Access-Control
DSO	Distribution-System-Operator
RSSI	Received-Signal-Strength-Indicator
EDs	End-Devices
PER	Packet-Error-Rate
CSS	Chirp-Spread-Spectrum
CoAP	Constrained-Application-Protocol
EUI	Extended-Unique-Identifier
OTAA	Over-The-AIR-Activation
ABP	Activation-BY-Personalization
ADR	Adaptive-Data-Rate
ToA	Time-on-Air
DH	Direct-Hop
NRH	Next-Ring-Hop
VH	Variable-Hop

Author details

Biswajit Paul[1*] and Rajesh Palit[2]

1 Electrical and Electronic Engineering (EEE), Shahjalal University of Science and Technology (SUST), Sylhet, Bangladesh

2 North South University, Dhaka, Bangladesh

*Address all correspondence to: biswajit-eee@sust.edu

References

[1] Leiva J, Palacios A, Aguado JA. Smart metering trends, implications and necessities: A policy review. Renewable and sustainable energy reviews. 2016 Mar 1;55:227-233. DOI: 10.1016/j.rser.2015.11.002

[2] Alahakoon D, Yu X. Smart electricity meter data intelligence for future energy systems: A survey. IEEE Transactions on Industrial Informatics. 2015 Mar 18;12(1):425-436. DOI: 10.1109/TII.2015.2414355

[3] Ancillotti E, Bruno R, Conti M. The role of the RPL routing protocol for smart grid communications. IEEE Communications Magazine. 2013 Jan 4;51(1):75-83. DOI: 10.1109/MCOM.2013.6400442

[4] Bou-Harb E, Fachkha C, Pourzandi M, Debbabi M, Assi C. Communication security for smart grid distribution networks. IEEE Communications Magazine. 2013 Jan 4;51(1):42-49. DOI: 10.1109/MCOM.2013.6400437

[5] Moslehi K, Kumar R. A reliability perspective of the smart grid. IEEE transactions on smart grid. 2010 May 27;1(1):57-64. DOI: 10.1109/TSG.2010.2046346

[6] Kumar P, Lin Y, Bai G, Paverd A, Dong JS, Martin A. Smart grid metering networks: A survey on security, privacy and open research issues. IEEE Communications Surveys & Tutorials. 2019 Feb 14;21(3):2886-2927. DOI: 10.1109/COMST.2019.2899354

[7] Uslar M, Rohjans S, Neureiter C, Pröstl Andrén F, Velasquez J, Steinbrink C, Efthymiou V, Migliavacca G, Horsmanheimo S, Brunner H, Strasser TI. Applying the smart grid architecture model for designing and validating system-of-systems in the power and energy domain: A European perspective. Energies. 2019 Jan;12(2):258. DOI: 10.3390/en12020258

[8] Ghasempour A. Internet of things in smart grid: Architecture, applications, services, key technologies, and challenges. Inventions. 2019 Mar;4(1):22. DOI: 10.3390/inventions4010022

[9] Nielsen JJ, Madueño GC, Pratas NK, Sørensen RB, Stefanovic C, Popovski P. What can wireless cellular technologies do about the upcoming smart metering traffic?. IEEE Communications Magazine. 2015 Sep 16;53(9):41-47. DOI: 10.1109/MCOM.2015.7263371

[10] Akpakwu GA, Silva BJ, Hancke GP, Abu-Mahfouz AM. A survey on 5G networks for the Internet of Things: Communication technologies and challenges. IEEE access. 2017 Dec 4;6:3619-3647. DOI: 10.1109/JIOT.2019.2948888

[11] Inga E, Cespedes S, Hincapie R, Cardenas CA. Scalable route map for advanced metering infrastructure based on optimal routing of wireless heterogeneous networks. IEEE Wireless Communications. 2017 Apr 24;24(2):26-33. DOI: 10.1109/MWC.2017.1600255

[12] Nandi S, Thota S, Nag A, Divyasukhananda S, Goswami P, Aravindakshan A, Rodriguez R, Mukherjee B. Computing for rural empowerment: enabled by last-mile telecommunications. IEEE Communications Magazine. 2016 Jun 23;54(6):102-109. DOI: 10.1109/MCOM.2016.7498095

[13] Bao L, Wei L, Jiang C, Miao W, Guo B, Li W, Cheng X, Liu R, Zou J. Coverage analysis on NB-IoT and LoRa in power wireless private network.

Procedia computer science. 2018 Jan 1; 131:1032-1038. DOI: 10.1016/j.procs.2018.04.252

[14] Raza U, Kulkarni P, Sooriyabandara M. Low power wide area networks: An overview. IEEE Communications Surveys & Tutorials. 2017 Jan 16;19(2):855-873. DOI: 10.1109/COMST.2017.2652320

[15] Rafi MN, Muaaz M. Performance Evaluation of the LoRa Protocol in the context of Smart Meter. arXiv preprint arXiv:1907.12355. 2019 Jul 4.

[16] Cano-Ortega A, Sánchez-Sutil F. Monitoring of the efficiency and conditions of induction motor operations by smart meter prototype based on a LoRa wireless network. Electronics. 2019 Sep;8(9):1040. DOI: 10.3390/electronics8091040

[17] Xia Z, Zhou H, Gu K, Yin B, Zeng Y, Xu M. Secure session key management scheme for meter-reading system based on lora technology. IEEE Access. 2018 Nov 27;6:75015-75024. DOI: 10.1109/ACCESS.2018.2883657

[18] Zacarias I, Gaspary LP, Kohl A, Fernandes RQ, Stocchero JM, de Freitas EP. Combining software-defined and delay-tolerant approaches in last-mile tactical edge networking. IEEE Communications Magazine. 2017 Oct 13;55(10):22-29. DOI: 10.1109/MCOM.2017.1700239

[19] Paul B. A Novel Energy-Efficient Routing Scheme for LoRa Networks. IEEE Sensors Journal. 2020 Mar 27;20(15):8858-8866. DOI: 10.1109/JSEN.2020.2983765

[20] Ye F, Qian Y, Hu RQ, Das SK. Reliable energy-efficient uplink transmission for neighborhood area networks in smart grid. IEEE Transactions on Smart Grid. 2015 Feb 5; 6(5):2179-2188. DOI: 10.1109/TSG.2015.2392130

[21] Komninos N, Philippou E, Pitsillides A. Survey in smart grid and smart home security: Issues, challenges and countermeasures. IEEE Communications Surveys & Tutorials. 2014 Apr 24;16(4): 1933-1954. DOI: 10.1109/COMST.2014.2320093

[22] Tan S, De D, Song WZ, Yang J, Das SK. Survey of security advances in smart grid: A data driven approach. IEEE Communications Surveys & Tutorials. 2016 Oct 11;19(1):397-422. DOI: 10.1109/COMST.2016.2616442

[23] He H, Yan J. Cyber-physical attacks and defences in the smart grid: a survey. IET Cyber-Physical Systems: Theory & Applications. 2016 Dec 1;1(1):13-27. DOI: 10.1049/iet-cps.2016.0019

[24] Stellios I, Kotzanikolaou P, Psarakis M, Alcaraz C, Lopez J. A survey of iot-enabled cyberattacks: Assessing attack paths to critical infrastructures and services. IEEE Communications Surveys & Tutorials. 2018 Jul 12;20(4): 3453-3495. DOI: 10.1109/COMST.2018.2855563

[25] Asghar MR, Dán G, Miorandi D, Chlamtac I. Smart meter data privacy: A survey. IEEE Communications Surveys & Tutorials. 2017 Jun 27;19(4): 2820-2835. DOI: 10.1109/COMST.2017.2720195

[26] Giaconi G, Gündüz D, Poor HV. Smart meter privacy with renewable energy and an energy storage device. IEEE Transactions on Information Forensics and Security. 2017 Aug 24;13(1):129-142. DOI: 10.1109/TIFS.2017.2744601

[27] Hong Y, Liu WM, Wang L. Privacy preserving smart meter streaming against information leakage of appliance status. IEEE transactions on information forensics and security. 2017 May 16;12(9):2227-2241. DOI: 10.1109/TIFS.2017.2704904

[28] Sun Y, Lampe L, Wong VW. Smart meter privacy: Exploiting the potential of household energy storage units. IEEE Internet of Things Journal. 2017 Nov 9;5 (1):69-78. DOI: 10.1109/ JIOT.2017.2771370

[29] Zhou J, Hu RQ, Qian Y. Scalable distributed communication architectures to support advanced metering infrastructure in smart grid. IEEE transactions on parallel and distributed systems. 2012 Jan 31;23(9): 1632-1642. DOI: 10.1109/TPDS.2012.53

[30] Ma S, Zhang H, Xing X. Scalability for Smart Infrastructure System in Smart Grid: A Survey. Wireless Personal Communications. 2018 Mar;99(1): 161-184. DOI: 10.1007/s11277-017-5045-y

[31] Tsado Y, Gamage KA, Adebisi B, Lund D, Rabie KM, Ikpehai A. Improving the reliability of optimised link state routing in a smart grid neighbour area network based wireless mesh network using multiple metrics. Energies. 2017 Mar;10(3):287. DOI: 10.3390/en10030287

[32] Saputro N, Akkaya K. Investigation of smart meter data reporting strategies for optimized performance in smart grid AMI networks. IEEE Internet of Things Journal. 2017 May 3;4(4): 894-904. DOI: 10.1109/JIOT.2017. 2701205

[33] Bera S, Misra S, Obaidat MS. Energy-efficient smart metering for green smart grid communication. In2014 IEEE Global Communications Conference 2014 Dec 8 (pp. 2466-2471). IEEE. DOI: 10.1109/ GLOCOM.2014.7037178

[34] Yang Z, Han R, Chen Y, Wang X. Green-RPL: An energy-efficient protocol for cognitive radio enabled AMI network in smart grid. IEEE Access. 2018 Mar 6;6:18335-18344. DOI: 10.1109/ACCESS.2018.2812191

[35] Anjana KR, Shaji RS. A review on the features and technologies for energy efficiency of smart grid. International Journal of Energy Research. 2018 Mar 10;42(3):936-952. DOI: 10.1002/er.3852

[36] Chen KL, Chen YR, Tsai YP, Chen N. A novel wireless multifunctional electronic current transformer based on ZigBee-based communication. IEEE Transactions on Smart Grid. 2016 Mar 22;8(4): 1888-1897. DOI: 10.1109/ TSG.2015.2510325

[37] Aalamifar F, Lampe L. Optimized WiMAX profile configuration for smart grid communications. IEEE Transactions on Smart Grid. 2016 Mar 14;8(6):2723-2732. DOI: 10.1109/ TSG.2016.2536145

[38] Khan AA, Rehmani MH, Reisslein M. Cognitive radio for smart grids: Survey of architectures, spectrum sensing mechanisms, and networking protocols. IEEE Communications Surveys & Tutorials. 2015 Sep 24;18(1): 860-898. DOI: 10.1109/ COMST.2015.2481722

[39] Meloni A, Atzori L. The role of satellite communications in the smart grid. IEEE Wireless Communications. 2017 Apr 24;24(2):50-56. DOI: 10.1109/ MWC.2017.1600251

[40] Paul B. Impacts of Multi-Hop Routing and Channel/Transmission Configuration Planning on LoRa Networks (Doctoral dissertation, University of Saskatchewan).

[41] Mikhaylov K, Petaejaejaervi J, Haenninen T. Analysis of capacity and scalability of the LoRa low power wide area network technology. InEuropean Wireless 2016; 22th European Wireless Conference 2016 May 18 (pp. 1-6). VDE. Print ISBN: 978-3-8007-4221-9

[42] Alsohaily A, Sousa E, Tenenbaum AJ, Maljevic I. LoRaWAN

radio interface analysis for North American frequency band operation. In2017 IEEE 28th Annual International Symposium on Personal, Indoor, and Mobile Radio Communications (PIMRC) 2017 Oct 8 (pp. 1-6). IEEE. DOI: 10.1109/PIMRC.2017.8292414

[43] Haxhibeqiri J, Van den Abeele F, Moerman I, Hoebeke J. LoRa scalability: A simulation model based on interference measurements. Sensors. 2017 Jun;17(6):1193. DOI: 10.3390/s17061193

[44] Mikhaylov K, Petäjäjärvi J, Janhunen J. On LoRaWAN scalability: Empirical evaluation of susceptibility to inter-network interference. In2017 European Conference on Networks and Communications (EuCNC) 2017 Jun 12 (pp. 1-6). IEEE. DOI: 10.1109/EuCNC.2017.7980757

[45] Liao CH, Zhu G, Kuwabara D, Suzuki M, Morikawa H. Multi-hop LoRa networks enabled by concurrent transmission. IEEE Access. 2017 Sep 22; 5:21430-21446. DOI: 10.1109/ACCESS.2017.2755858

[46] Lam KH, Cheung CC, Lee WC. Rssi-based lora localization systems for large-scale indoor and outdoor environments. IEEE Transactions on Vehicular Technology. 2019 Sep 9;68(12):11778-11791. DOI: 10.1109/TVT.2019.2940272

[47] Lemic F, Behboodi A, Famaey J, Mathar R. Location-based discovery and vertical handover in heterogeneous low-power wide-area networks. IEEE Internet of Things Journal. 2019 Aug 16; 6(6):10150-10165. DOI: 10.1109/JIOT.2019.2935804

[48] de Castro Tomé M, Nardelli PH, Alves H. Long-range low-power wireless networks and sampling strategies in electricity metering. IEEE Transactions on Industrial Electronics. 2018 Mar 15;66(2):1629-1637.

[49] Rizzi M, Ferrari P, Flammini A, Sisinni E. Evaluation of the IoT LoRaWAN solution for distributed measurement applications. IEEE Transactions on Instrumentation and Measurement. 2017 Sep 13;66(12): 3340-3349.

[50] Paul B. A Novel Mathematical Model to Evaluate the Impact of Packet Retransmissions in LoRaWAN. IEEE Sensors Letters. 2020 Apr 13;4(5):1-4. DOI: 10.1109/LSENS.2020.2986794

[51] Aref M, Sikora A. Free space range measurements with Semtech LoRa™ technology. In2014 2nd International Symposium on Wireless Systems within the Conferences on Intelligent Data Acquisition and Advanced Computing Systems 2014 Sep 11 (pp. 19-23). IEEE. DOI: 10.1109/IDAACS-SWS.2014.6954616

[52] Paul B, "An Overview of LoRaWAN", WSEAS TRANSACTIONS ON COMMUNICATIONS, vol. 19, pp. 231, 2021. DOI: 10.37394/23204.2020.19.27

[53] Paul B, Matin MA. Optimal geometrical sink location estimation for two-tiered wireless sensor networks. IET wireless sensor systems. 2011 Jun 1; 1(2):74-84. DOI: 10.1049/iet-wss.2010.0073

Chapter 4

Assessment of Power Plants in the Western Region of Libya during a Period of Insecurity

Khaled Mabrouk Alkar

Abstract

After the uprising in Libya in 2011, several outages and blackouts occurred in the electrical grid. The western region of Libya is the most affected part especially after the civil war in Tripoli 2014. This chapter focus on the assessment of energy production by Al-Zawia Combined Cycle Power Plant "Al-Zawia CCPP" and Western Mountain Power Plant during the period of blackouts and insecurity. In addition, to figure out the main causes of the frequent blackouts and outages in order to find practical solutions to ease the severity of the problem. This research is done based on the data are collected from the recorded data in Al-Zawia CCPP and Western Mountain power plant during the last two years 2019–2020. The data shows the instability of the annual energy produced from Al- Zawiya PPCC and Western Mountain Power Plant in 2019, also illustrates the improvement in the total produced energy by the six gas units of the Western Mountain power plant after the end of the war on Tripoli in 2020. However, the data shows the deficiency of the Western Mountain power plant to operate at its full capacity, especially in August 2020 due to the lack of maintenance.

Keywords: Assessment, blackout, Western Mountain power plant, Al-Zawia CCPP, outages, western region of Libya

1. Introduction

Power plants, transformers, transmission lines, and distribution stations are the main parts of the power system network [1]. Failure in any part of those could lead to blackout or outages. The European Network of Transmission System Operators expresses the blackout as "the interruption of electricity generation, transmission, distribution, and consumption processes, when operation of transmission system or a part is terminated" [2, 3]. In the last decade, the Libyan power grid faced numerous blackouts and outages and the consequences of these blackouts and outages were costly [4]. Therefore, more studies have to be done to reduce the effect of these problems. Some reasons for the continuous interruptions and load shedding are; the armed conflicts, postpone of overhaul units of the power plants. In this chapter, field study on the causes of power outages and blackout in Western Mountain power plant & Al-Zawia CCPP is presented. As well as, an assessment of the generation power units in both power plants during the period of the outages and blackouts 2019–2020. In order to suggest practical solutions.

IntechOpen

1.1 Research questions

- What are the main reasons of the current outages?
- Does the GECOL," the General Electricity Company of Libya power system" able to solve these problems?
- Is there an improvement in the energy supply to the consumers in 2021 or will it be worse?
- Is there a possibility for the electricity company to invest in alternative energies such as the solar and wind energy to cover the deficit in energy production?
- Is there a possibility to reconnect the network with the countries of the Maghreb Arabic and the European Union, or are there technical reasons that prevent this?
- to what extent it is possible to address these issues or find effective solutions to them to alleviate the suffering of the citizen in the difficult conditions of the country?

1.2 Relevance and important of research

The idea of the research is to investigate the real causes of the blackouts and frequent outages in energy supply to consumers in Western region of Libyan power system grid. It is possible to achieve this goal by analyzing the performance of the power plants in Western region of Libya. Which are Alzawia Combined Cycle Power Plant" Alzawia CCPP" with a total installed capacity of 1440 MW and Western Mountain power plant with a total installed capacity of 600 MW [4]. These power plants represent more than a quarter of the energy produced in Libya. Therefore, focusing the study on these stations may help find suitable solutions for the electric company.

1.3 Literature review

The major industrial continents of North America, Europe, Asia, and Australia have faced the same causes of blackouts since 1965 [1]. In September 2003, due to cascade tripping in the transmission line (380 kV 220) linking Italy and northern Europe, which resulted in the largest blackout that Italy has witnessed, affecting more than 55 million people [2]. In 2010, there were many blackouts in the supply of electrical energy to consumers, which caused millions of customers to lose energy for long hours. September 2011, due to a blackout in the southwestern ocean, which led to a loss of energy for more than 12 hours, and about 2.7 million people have affected in some southwestern states in North America, such as residents of San Diego, California, Arizona, and Mexico. The main reason for the occurrence of blackout is an increase in the load during the peak period on a major transmission line, which caused the collapse of the electrical network system [3]. In 2019, there was a study on the performance of the Zawia CCPP during the blackouts and normal operation conditions [4]. Therefore, when a blackout or power outage takes place, the consequences can be costly. It can cause the loss of life of intensive care patients and children where childbirth needs a nursery, in addition to financial and industrial losses. There are several reasons that lead to blackout and power outages in power system grid such as; faults in transmission lines, increase of load demand, failure in protection system, poor maintenance of the equipment, human error, over frequency, cyber-attacks, voltage fluctuation, a short circuit, lighting strike, severe weather, and ice storms [3, 5].

2. Causes of outages and blackouts in Libya

Since the 2011 uprising, the Libyan power grid has suffered from several operating and infrastructure problems. Most of the maintenance projects have stopped due to security and political problems of the country. Among the most prominent of these problems are; the repeated attacks on GECOL's assets and on workers, stealing of electrical equipment such as copper wires, transformers, and electrical transmission towers. Which caused the GECOL to be unable to perform some of the necessary periodic maintenance for most of the power plants. These problems resulted in a decline in the performance of the electricity network and a severe deficiency in the production capacity of the generating stations, long outages, and blackouts in most areas in the country [6].

2.1 A brief overview of power plants in Libya

The GECOL owns about 26 electric power plants. These power plants contain 85 generating units of different ages, sizes, and operating technology. These units are mostly concentrated on the sea in the North [6]. The official installed capacity by the GECOL until 2017 is 10.238 GW, while the energy available to consumers only 5.53 GW until the date of this study, which represents 52% of the total capacity of the GECOL. Besides, 19 units of the GECOL units are must be discontinued due to the end of their useful life and the futility of their continued operation [6]. In short that, there is a deficit in the power production of the GECOL at a rate of approximately 25% of the average production, compared to the maximum value of the energy demand, which is 7.5 GW [7].

2.2 Overview of control department in GECOL

The control infrastructure in the GECOL consists of several levels. At the top of the system, is the National Control Center (NCC), as shown in **Figure 1**. Its mission

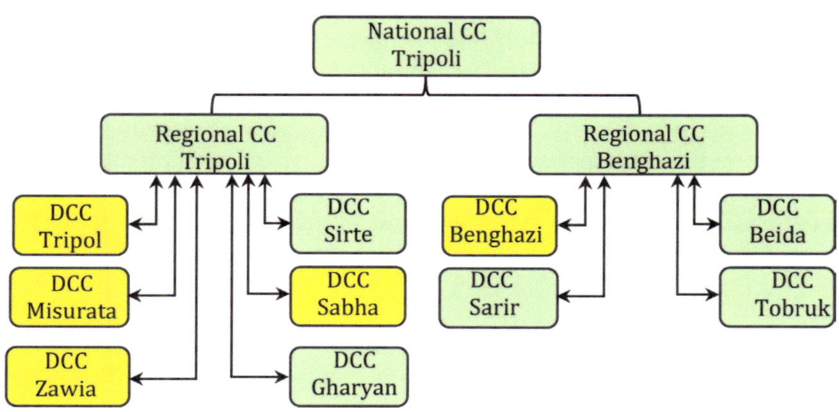

Figure 1.
Distribution control system of GECOL precedence DCC's in yellow [6].

is to supervise, coordinate and control the power plants. The main transmission network is 400 KV, whereas 220 KV is the subnet that linking between Egypt in the east and Tunisia in the west [6].

Libya's control system is divided into two parts: the Tripoli Regional Control Center (TRCC) in the Western part and the Benghazi Regional Control Center (BRCC) in the East, both parts responsible for controlling and operating the substation 220 kV. In addition, the Distribution Control Center is responsible for the medium and low voltage network lines of 66 kV, 30 kV and 11 kV. Whereas NCC, TRCC and BRCC are supervised by the General Control Department. The Distribution Control Center (DCC) is under the control of the General Distribution department and the Medium Voltage General Department. These control centers are linked to the generation stations and substations and are controlled by a fiber optic system. This system consists of ground cables, Optic Ground Wires (OPGW) transmission towers, and some old power carriers and microwave connections [6].

After the uprising in 2011, the transmission lines were mostly damaged due to the civil war, which results in losing communication links one by one and a huge loss in each fiber optic-based data and voice communications links between the power plants and network substation.

The mentioned problems in the communication links and some Remote Terminal Units (RTUs) in the stations, led to a loss of 80% of the data compared to the year 2011, where there is only a 2% loss. The NCC and TRCC computer system only controls 20% of the data that covers main points in the network, which provides a partial picture to the control engineers [6]. Some lost data is covered by Supervisory Control and Data Acquisition (SCADA) and the rest of the data is manually entered to give a fair picture of the network. Consequence, NCC is completely dependent on the operators in all stations and on voice communications linked with the operators to effectively control the network [6].

Through investigations and studies for the GECOL Control General Department, it was established that the four causes of the blackouts in 2017 were for the following reasons; firstly, the interruption of some transmission lines has shaking the operative reliability of the electrical power grid. Secondly, outages in some transmission lines, which weakened the operational capacity of GECOL power system. Thirdly, all blackouts were due to in faults in the transmission network and as a result the loss of some generation units. Fourth, the southern region suffers from voltage instability, which caused the network frequency to rise, followed by a transmission failure, following the initial transmission fault clearance. Finally, the recorded data of frequency and voltage response after the fault is removed, indicates the possibility of a "Fault-induced delayed voltage recovery" (FIDVR), causing an increase in frequency and loss of generation. This phenomenon has been observed to occur in the electrical power grid with large loads of inductive motors, specifically the loads of air conditioners. The Libyan electrical power grid is full and saturated with air conditioner loads, which is easy to assure that at the end of the summer and with temperate temperatures, particularly in a two-week period of September 2017, peak loads decreased by 2 GW because consumers did not need to use their air conditioners. So there is no sure solution to this problem, which requires more research and studies to solve the problem and avoid the consequences resulting from it [6].

3. The construction of Al-Zawia combined cycle power plant and its connection with the power system grid

Al-Zawia CCPP contains of three steam turbines and 6 gas turbines, beside two auxiliary motor turbines. The six gas turbines are known as (GT11, GT12, GT13,

GT14, GT15 and GT16), whereas, the steam turbines are called (ST10, ST20 and ST30). The auxiliary motor turbines named (TM05 & TM10). The nine units of steam and gas are divided to three compounds, each compound contains two gas turbines coupled with steam turbine [4].

Al-Zawia CCPP is connected to the general power system grid through eight electric circuits to transfer the produced energy. There are two main transmission lines 400 KV and the rest are 220 KV. Four transmission lines are linked to Western Tripoli power plant. The 400 KV transmission lines need to reduce the transferred energy to 220 KV using step-down transformer 400KV/220 KV, and the other two 220 KV transmission lines are linked directly to the Bus-Western Tripoli 1 &2 respectively. The fifth and sixth transmission lines 220 KV are connected to Al-Harsha substation through Bus Al-Harsha 1 &2. The last two transmission lines 220 KV are joined with Zahra Gas Turbine power plant.

From **Figure 2**. it can be noticed that the annual energy produced from Al-Zawiya PPCC for the year 2019 is fluctuating and not constant in all months of the year. The best performance for Al- Zawiya PPCC was in the winter season, specifically in January, where it reached 884274 MWh, then it declined by 19.30% In February, where it reached 731551 MWh. August 2019 was the most difficult period of the year for the power plant in a matter of production. Due to the increased energy demand, in the summer season, with the rise of temperatures, the power plant production decreases as well as its efficiency by 55.3% compared to January. Only 394,491 MWh were produced by the power plant during this month which means a very sharp drop in the power plant's production, which is expressed by the plant's inability to produce steadily and continuously. One of the main reasons that led to deficiency production of power plant in August, the worst at all, is the frequent blackouts. There were eight blackouts during this month, in particular, from 02 to 08 to 12–08-201. 4th -August was the worst day ever, where three blackouts took place within a 24-hour. In the following figures (**Figures 3–5**), there are more details on the power plant's performance in some of these days [8].

The daily report of Al-Zawia CCPP shows the status of the nine units on the day of blackout, as following; the GT11 and GT12 were tripped at 10:17 on 02/08/2019 due to a shake in the grid. The GT11, GT13, and GT12 were restored at 12:51, 13:16, and 14:05 consequently. The GT13 tripped at 14:07 due to a shake in the grid and it was restored at 14:29, then, it tripped again at 15:39 for the same previous reasons. The GT12 tripped again at 15:39 due to a shake in the grid. The

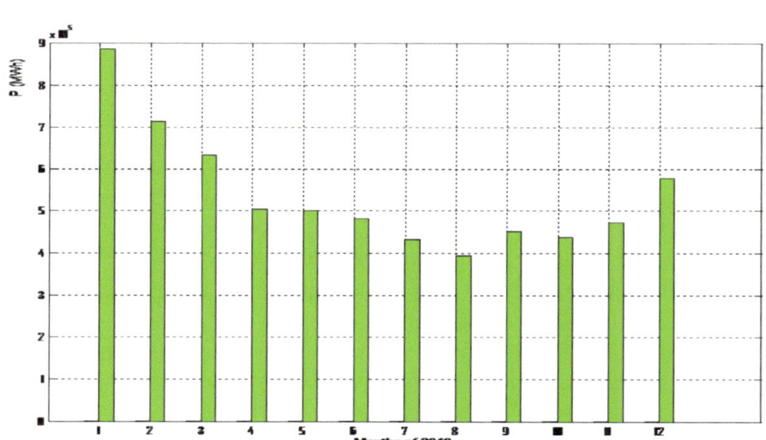

Figure 2.
The total produced energy by Al-Zawia CCPP in 2019.

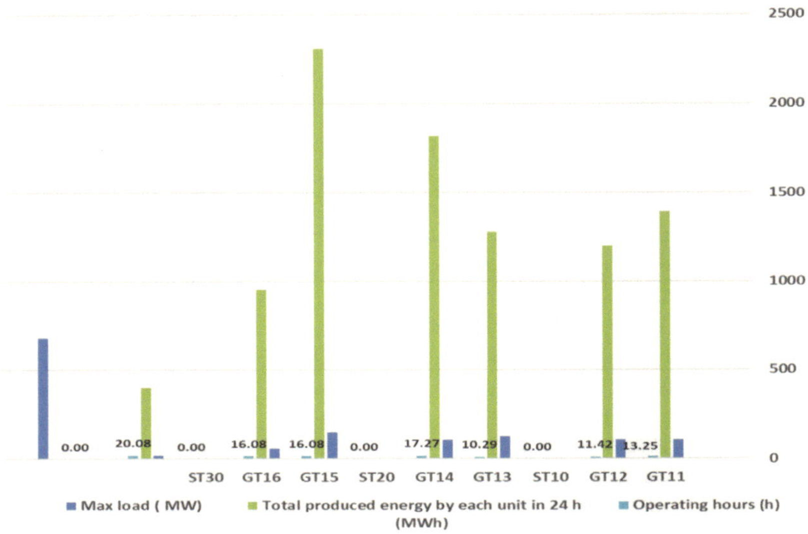

Figure 3.
The produced energy by Al-Zawia CCPP on 02-08-2019.

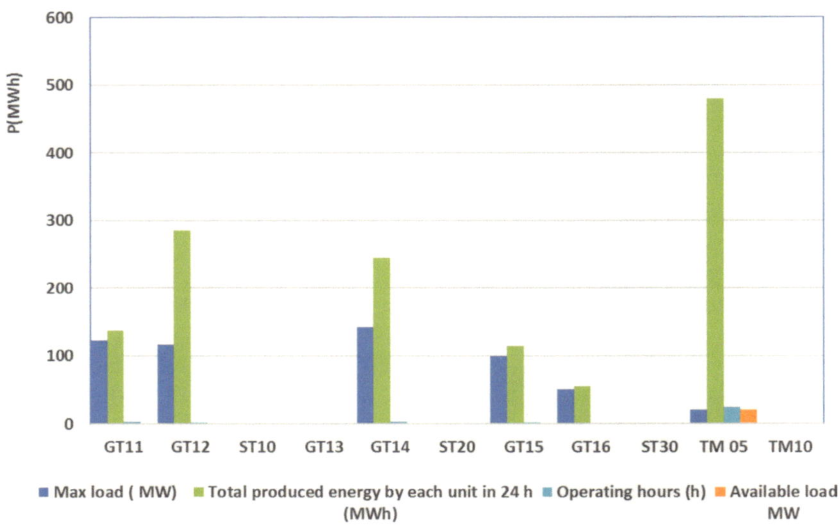

Figure 4.
The statu of the nine units on the day of blackout 03-08-2019.

GT11, GT15, and GT16 tripped because of a shake in the grid, which let to blackout at 16:52. TM 05 and the GT14 were restored at 21:00, 21:59 respectively. Once again, a complete Blackout At 22:34, and TM 05 stayed disconnected.

As result of the repeated blackouts, it can be shown in **Figures 3–6** the sharp decline of the daily produced energy by the six gas units, and not to mention that the three steam units were all out of service.

On the day after 03-08-2019, the scenario of the blackout continued until, the gas units were started up one by one as following; GT14 at 18:12, GT15 at 18: 18, GT11 at 18:35, GT16 at 18:56, and GT12 at 19:38. On the day after 03-08-2019, the scenario of the blackout continued until, the gas units were started up one by one as following; GT14 at 18:12, GT15 at 18: 18, GT11

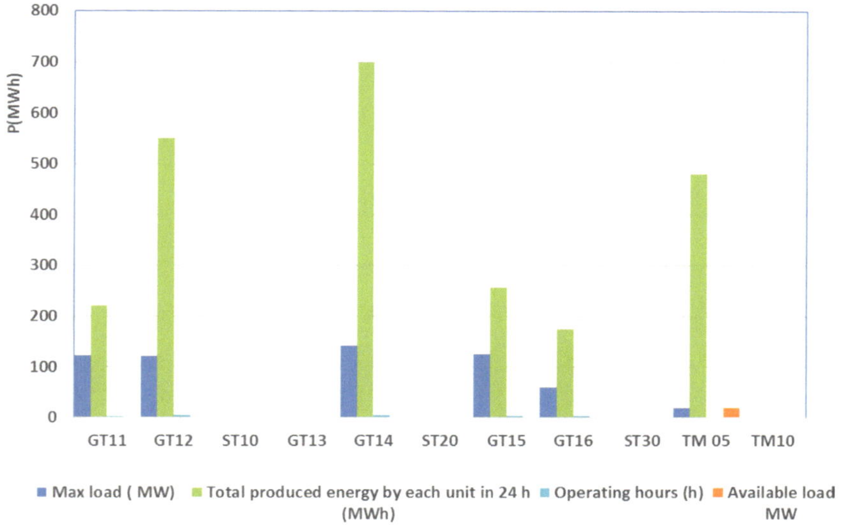

Figure 5.
The produced energy by the units of Al-Zawia CCPP on of the day of the blackout 4-8-2019.

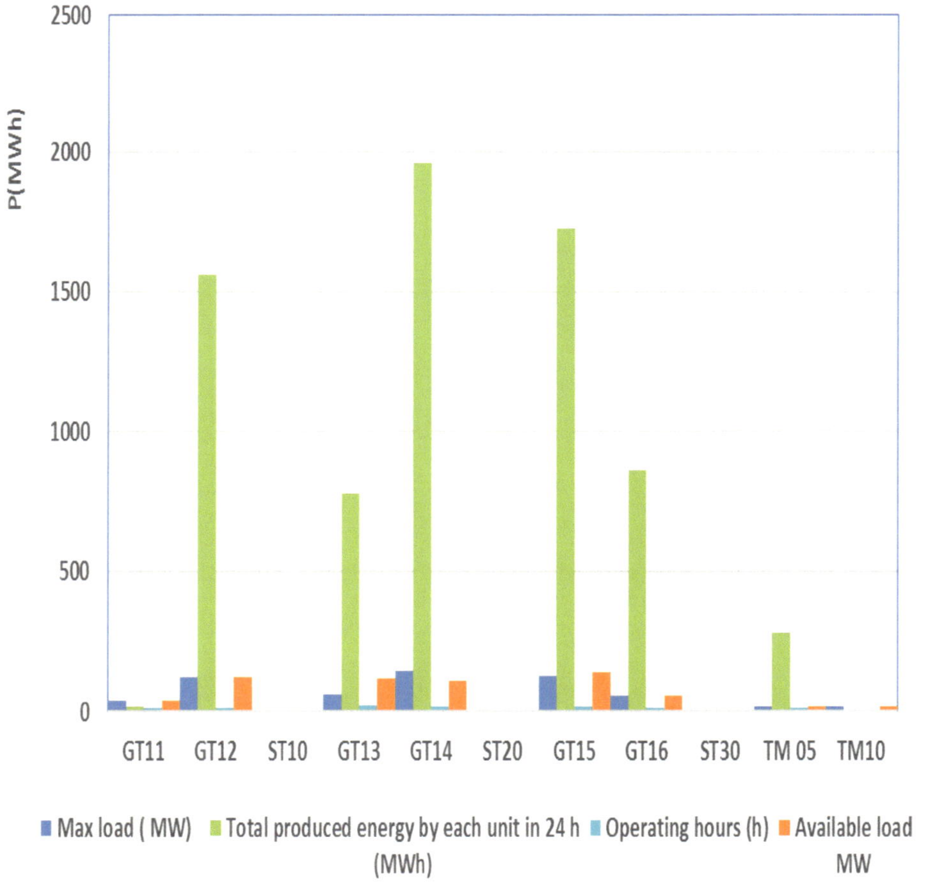

Figure 6.
The performance of gas units of Al-Zawia CCPP on one day of the blackouts 07-08-2019.

at 18:35, GT16 at 18:56, and GT12 at 19:38. At the same day, at 20:20 another blackout was occurred. The gas units were started up in order GT12 22:08, GT11 at 22:43, GT14 at 23:04, GT15 23:59 and GT16 00:17. In the third day in row,

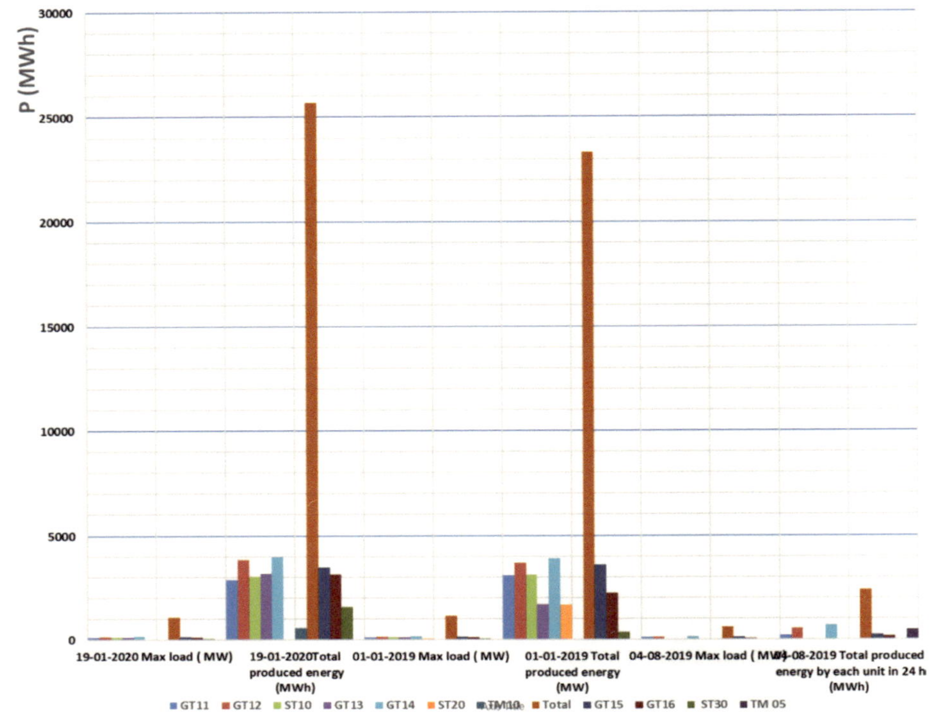

Figure 7.
Comparative of the total produced energy on normal operation day and on the days of the blackouts.

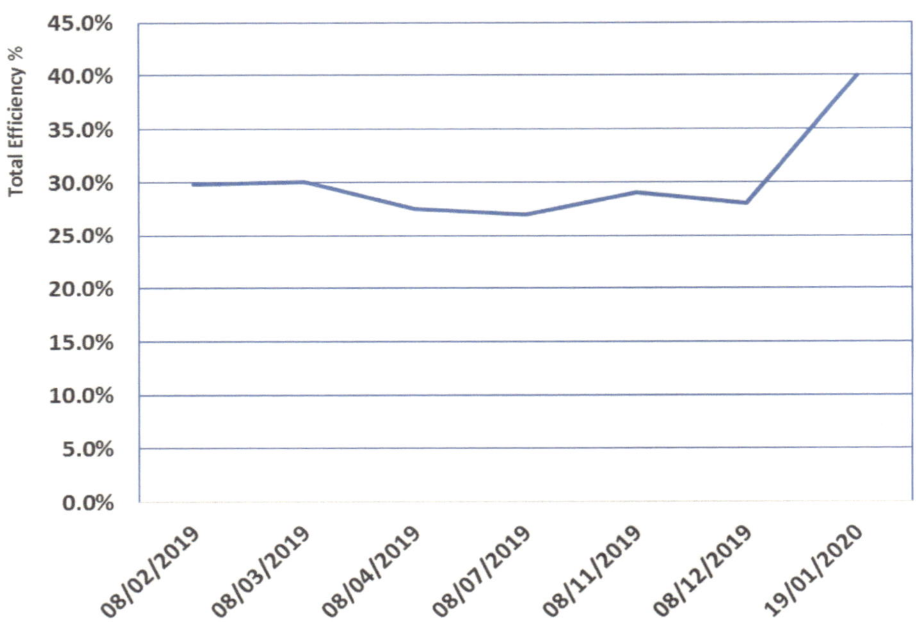

Figure 8.
Comparative of the total efficiency on normal operation day and on the days of the blackouts.

on 04-08-2019 at 02:32 all gas units were shut down again which means blackout again.

After a long exhausting night of continuous and tedious work from the plant's engineers and operators to restore the units to the network, they succeeded to restore four out of six respectively, GT14 at 11:45, GT12 at 12:10, GT15 at 12:30 and GT16 at 12:50 on 04-08-2019. After 1.16 h, the restored units went blackout again. In the evening, the four gas units were successfully restored within 1.10 h, as following, GT12 at 20:45, GT14 at 20:55, the GT16 at 21:30, and GT15 at 21:55″. Within 24 hours, the third blackout occurred at 22:28. On 05-08-2019, On 05-08-2019, the swinging four units were restored and connected to the power system grid again in order, the GT16 at 01:05, GT14 at 01:15, GT15 at 1:37, and GT12 at 03:40. In the morning at 09:07, a blackout took place again.

On 07-08-2019 at 13:56 another blackout occurred. 03:31 hours later, four gas units were restored in sequence GT14 at 17:25, GT15 at 19:08, GT16 at 17:34, and GT11 at18:06.

From next **Figures** 7 and **8** it can be noticed the following; the increase of the daily total produced energy by Al-Zawia CCPP at the beginning of the year 2020 compared to the year before. The sharp decrease in operating efficiency in the days of blackouts compared to the normal operation days.

4. Western Mountain power plant

Western Mountain Power Plant (Ruwais) is located southwest of the capital Tripoli, with an estimated distance of about 250 km. It depends on natural gas to produce electricity as the main fuel, and on diesel as fuel in case of emergency. It consists of four units as a basic project and two units as an expansion project, meaning the total is six units, each unit produces 156 MW, which means 936 MW total. There is a Central Control Room (CCR) through which all units can be controlled, and each unit can also be controlled through Power Control Center (PCC) sub-controller.

For the connection with GECOL power system, there are six units (GT11, GT12, GT13, GT14, GT15 and GT16). The first four units GT1- GT4 are connected to transmission lines 220 KV through step-up transformers. There are five transmission lines 220 KV, the first and second lines are connected to Shakshuk substation and known as "Shakshuk1 and Shakshuk2". The third and fourth lines are linked to Al-Rabeta, Zahra substations respectively, and the last line is connected to Tataouine – Tunisia photovoltaic 10 MW. The fifth and sixth units (GT5, GT6) are connected to a 400 kV high voltage line via step-up transformer and linked with Ghadames substation.

Figure 9 shows the total produced energy from natural gas by the six units of the Western Mountain power plant, the consumed energy from natural gas, and the total available loads in 2019. It can be seen the variation of the six units in the amount of annual production energy. It can be noticed that the sixth and second units are considered the best in terms of performance. They produced 945001.25 MW and 861 103.80 MW respectively. While the fourth and first units are the worst in terms of the amount of electrical energy production, the reasons behind are; the units exceeded the equivalent operating hours for overhaul and that the last one for GT14 was on 5 July 2016, as well as, the number of scheduled stopping hours were 2288.23 hours as shown in **Table 1**. As for the first unit GT11, the last overhaul was on 23 November 2015, and the hours of sudden stop were 2990.6 hours, which explains the reason for the low amount of energy production compared to other units [9].

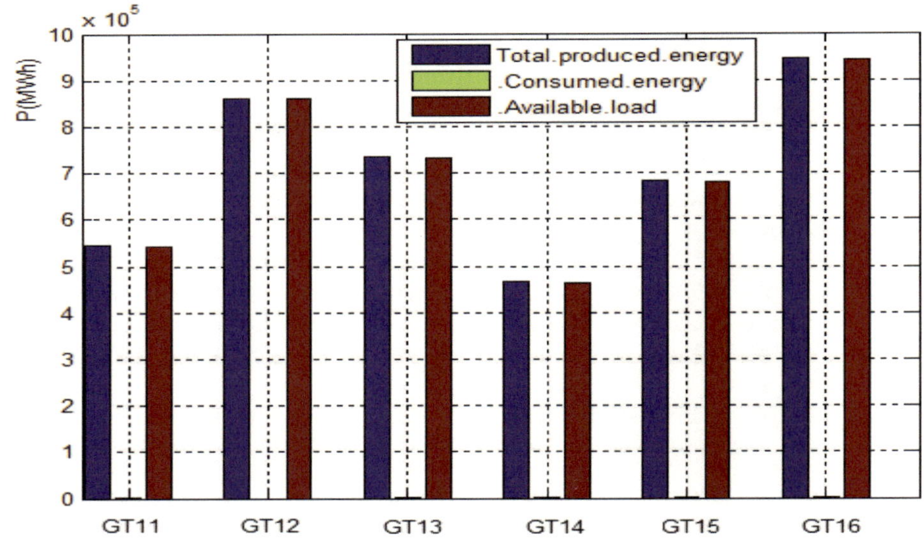

Figure 9.
The total produced energy by the six units of the Western Mountain power plant in 2019.

Data	GT11	GT12	GT13	GT14	GT15	GT16
Number of actual operating hours (h)	5598.02	8399.9	8282.71	6447.01	8164.76	8094.43
The number of equivalent operating hours (h)	6524.64	30156.76	9046.76	195511.86	9080.38	9080.8
The number of scheduled stoppage hours	171.37	226.48	415.01	2288.23	475.3	571.99
The number of hours of forced stoppage	0	0	0	0	0	0
The number of hours sudden stoppage	2990.6	133.61	62.28	24.76	120.03	93.58
The number of run times	16	25	15	17	19	11

Table 1.
Annual report on the number of operating hours of the Western Mountain power plant in 2019.

Figure 10 shows the total energy produced from the six units of the Western Mountain power plant for the year 2020. There is a significant increase for production of the GT11 compared to the year 2019, which reached 35% from its production. The same applies to GT14 and GT15 whose performance improves by 21.5% and 20.01%, respectively. There is a slight decrease in the production quantity of the GT16. The rise of energy production by GT11 goes to the increase of the number of actual operating hours from 5598.02 in 2019 to 8394.45 in 2020, as well as the decrease in the number of scheduled stoppage hours from 171.1 in 2019 to 67.11 in 2020, as shown in **Tables 1** and **2**.

Figure 11 shows the significant improvement in the performance of the Western Mountain power plant through the increase in energy production and the amount of

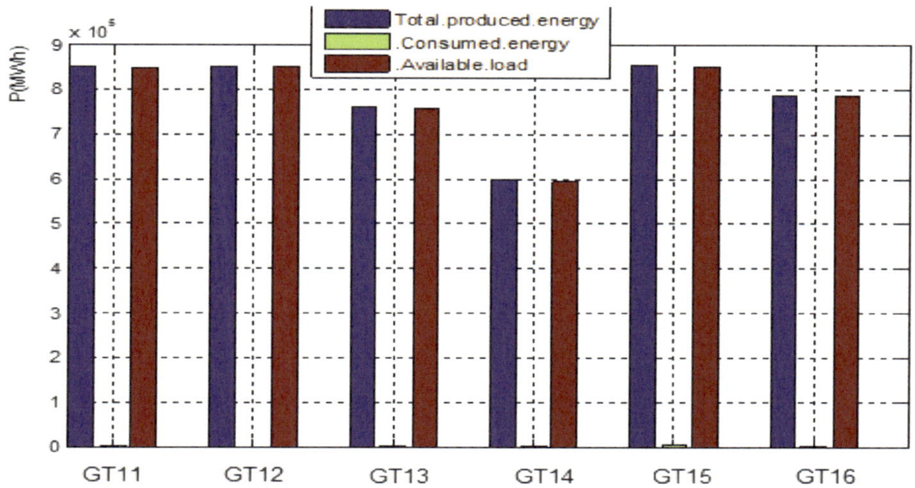

Figure 10.
The total produced energy by the six units of the Western Mountain power plant in 2020.

Data	GT11	GT12	GT13	GT14	GT15	GT16
Number of actual operating hours (h)	8394.45	8602.6	8541.78	8464.1	8474.25	6208.71
The number of equivalent operating hours (h)	12082.2	12336.03	11443.46	9364.01	9957.83	8205.9
The number of scheduled stoppage hours	67.11	209.84	103.24	210.39	225.77	473.83
The number of hours of forced stoppage	0	0	0	0	0	0
The number of hours sudden stoppage	322.44	69.69	139.02	109.51	83.98	2100.9
The number of run times	35	44	32	23	26	24

Table 2.
Annual report on the number of operating hours of the Western Mountain power plant in 2020.

energy sent to consumers in 2020 compared to the year before. This improvement has achieved because of; in the first quarter of 2020, the civil war stopped in the capital, Tripoli, where the National Control Center Tripoli of GECOL, some main transmission lines, substations, and Tripoli South Power plant are on the armed conflict area "south of Tripoli". This puts this equipment under the risk of destruction, damage, and cutting off some main transmission lines. As result, these risks effect on the stability of the power system grid and cause of frequent outages and blackouts.

Due to the high temperatures in summer season, specifically in August, which causes an increase of electricity demand by customers. Those conditions, put the power plants under the pressure to cover the increase of energy demand. This month was chosen to study the performance of the six units of the Western Mountain power plant under these conditions. **Figures 12–17** show the performance of the six units GT11, GT12, GT13, GT14, GT15 and TG16 in terms of maximum, minimum and average loads.

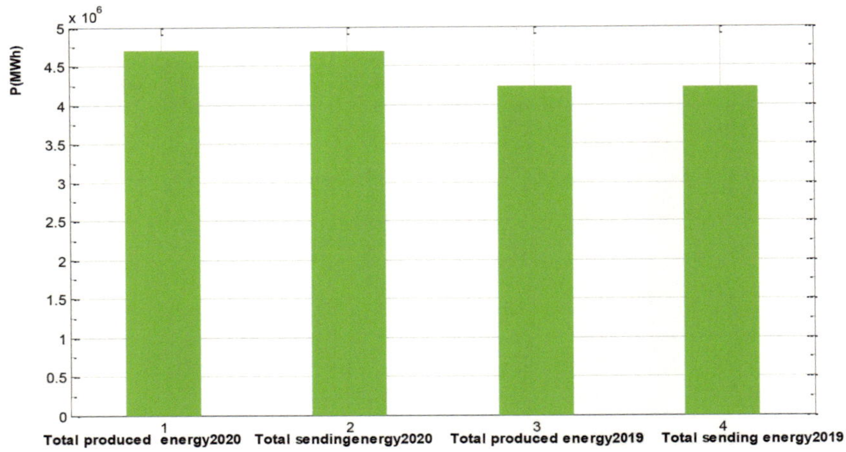

Figure 11.
Comparative of Western Mountain power plant production between 2019 & 2020.

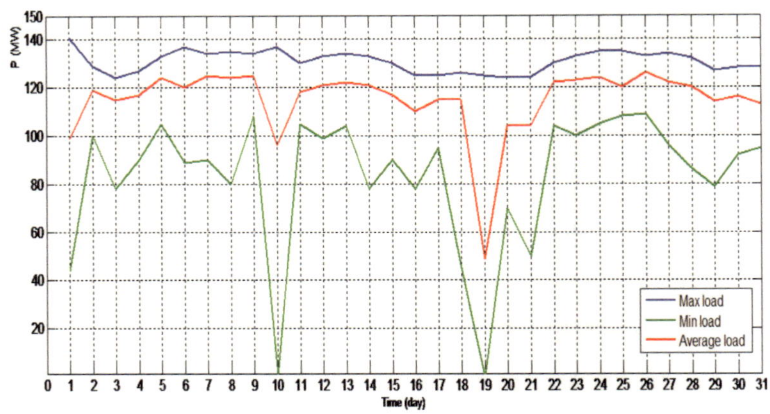

Figure 12.
The performance of GT11 in Aug-2020.

From **Figure 12**, it can be seen that the minimum loads in GT11 decreased sharply twice in August, the first one was on 10-Aug-2020 due to over frequency as reported on the recorded data, which led to the trip of GT11 0.5 h. The second sharp decline was on 19-Aug-2020 for the same reason and caused an outage of 1.9 h.

The minimum loads of GT12 decline suddenly to zero four times as shown in **Figure 13**. Which means four outages, the first sudden decline was on 03-August –2020 because of a trip of GT12 due to surge protection. The second and third decline was on 10 & 19 -Oct led to a trip of GT12 due to over frequency. Tripped of GT12 on 29- Aug due to the high temperature of the turbine outlet caused the last sharp decline.

There were sharp decreases in the remaining four units GT13 and GT16, as shown in **Figures 14–17**. The units GT13 &GT15 declined to zero minimum loads three times in Aug- 2020, which mean three interruptions, while the GT14 was the worst among the six units s in terms of performance and quantity of production. The minimum loads of GT14 decreased to zero six times in a period 9–26 Aug, which shows the clear effect of lack of maintenance. The GT16 was the best performance compared to the other units even, the minimum loads reduced zero twice in the same period.

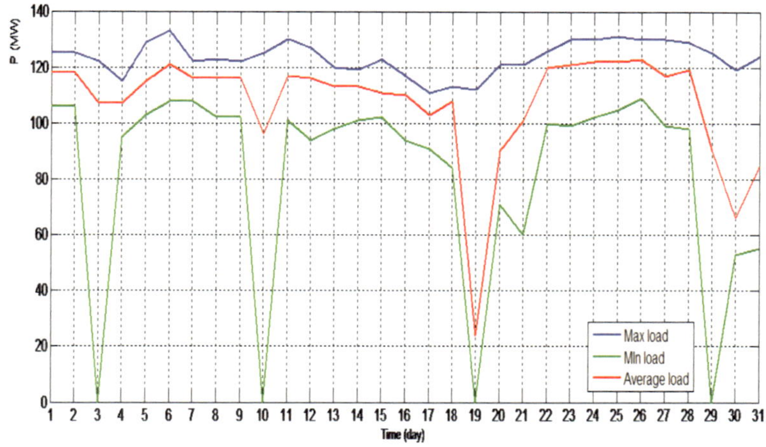

Figure 13.
The performance of GT12 in Aug-2020.

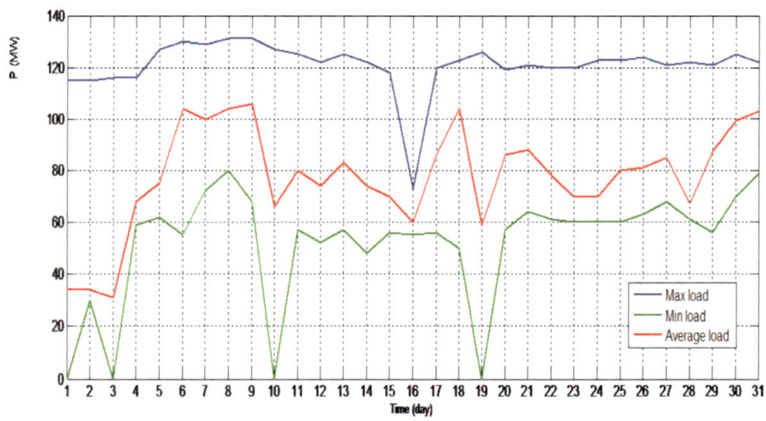

Figure 14.
The performance of GT13 in Aug-2020.

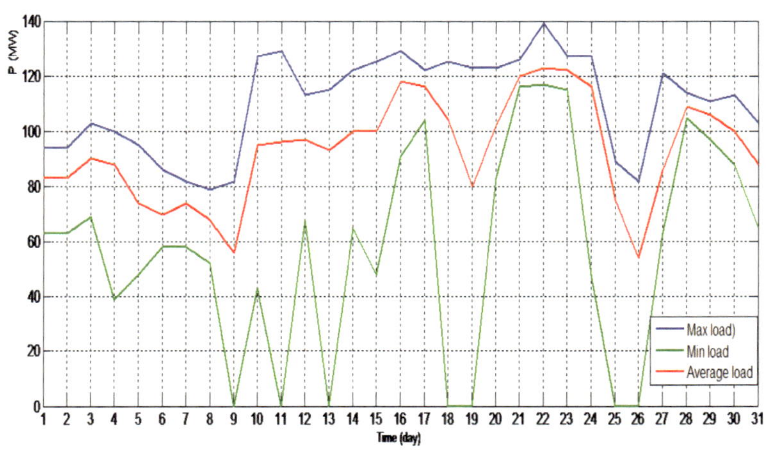

Figure 15.
The performance of GT14 in Aug-2020.

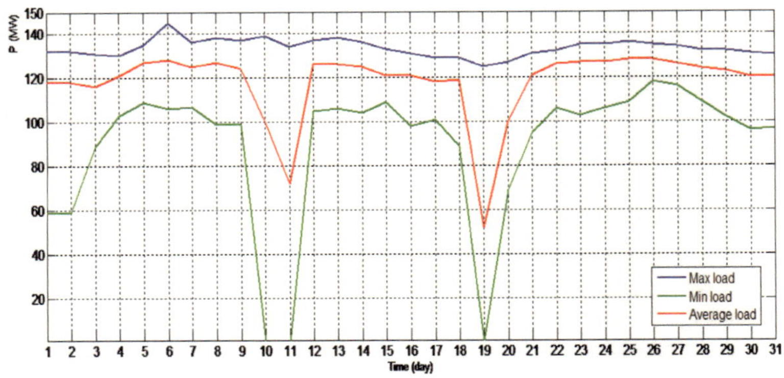

Figure 16.
The performance of GT15 in Aug-2020.

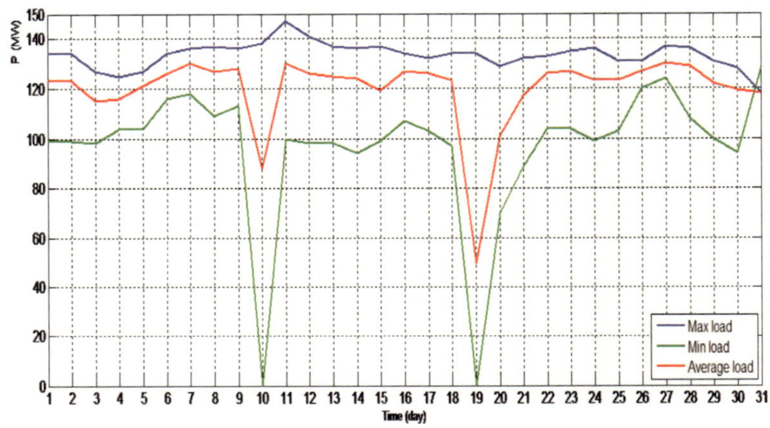

Figure 17.
The performance of GT16 in Aug-2020.

The cause of interruptions in GT14 is due to the following: four interruptions to increased frequency, the fifth programmed for maintenance, and the sixth interruption because of repairing air leakage. The units GT15 &GT16 were tripped twice on 10 &19 – Aug, due to over frequency. From all **Figures 12–17**, it can be noticed that all units were tripped on 10 & 19 -Aug, except, GT14 due to over frequency and shake in the power system grid.

5. Conclusion

In this chapter, outages and blackouts that occurred in the general electrical grid in 2019–2020 have been discussed in detail as follows; from the point of the annual energy produced by Al- Zawiya PPCC in 2019 and during the period of blackouts 02–07 Aug –2019. Also, Comparative of the annual energy produced by Western Mountain Power Plant in 2019 & 2020. In addition, an assessment of the performance of the six gas units of Western Mountain Power Plant during the peak load demand on Aug-2020. From the presented data, the main causes of outages and blackouts are as follows; the inability of the two power plants to operate at their full capacity during the peak load to cover the load demand, the exit of foreign experts for security reasons caused the delay in the maintenance and overhaul of the units, the deliberate sabotage of the control room of the Western power plant by armed militias, the division of the power system grid into several parts as well as the loss of the main transmission line 400 KV that linked the Western region to the East is leaving the network very weak, the NCC and TRCC computer systems only control 20% of the data that covers the main points in the network, all faults in the transmission network led to the loss of some generation units. The SCADA only covers some lost data and the rest is manually entered to give a fair picture of the network. Finally, the civil war and political division make the maintenance work very difficult to repair the broken equipment near the conflict zone.

The proposed solutions to overcome the outages and blackout are protecting power plants, transmission lines, and the equipment of the electric company from theft and deliberate destruction. Moreover, reconnecting the main transmission line 400 kV that linking East region with the West, re-activate the automatic operation of the National Control Center for remote generation stations, update Scada system, maintenance of protection relays for the main transmission lines and activation of the deactivated automatic connection and disconnection system. Furthermore, maintenance of all broken generation units and overhaul of the operating units on time, reconnecting the main transmission line 400 KV linking between Tunisia, Libya, and Egypt to increase the stability of the network. Finally, in war zones and conflict areas, where the transmission lines power system equipment is under the threat of deliberate destructions and stealing, using renewable energies, smart grid and smart metering to supply the customers with power and to compensate the voltage drop for increasing network stability are the best solutions.

Author details

Khaled Mabrouk Alkar
Higher Institute of Science and Technology, Subratha, Libya

*Address all correspondence to: kalkar76@gmail.com

References

[1] Liu S. Analyses and Discussions of the Blackout in Indian Power Grid Analyses and Discussions of the Blackout in Indian Power Grid. 2018;(February).

[2] Prasad S, Prasad S, Sourab M, Hembrom S. Blackout: Its Causes and its Prevention. 2016;4(02):2-3.

[3] Alhelou HH, Hamedani-Golshan ME, Njenda TC, Siano P. A survey on power system blackout and cascading events: Research motivations and challenges. Energies. 2019;12(4):1-28.

[4] Alkar KM, Meto M, Jamjum M, Amar K. Performance Evaluation of the Blackout and Power Outages in Libyan Power Grid-Al-Zawia Combined Cycle Power Plant Case Study. Proc - 2019 1st Int Conf Sustain Renew Energy Syst Appl ICSRESA 2019. 2019;

[5] Shivashimpiger S, Siridevi NC, Parthasarathy V, Ravikumar HM. A Qualitative analysis of Power System Blackouts -Northern Grid Case Study. 2015;3(6):373-379.

[6] Gecol I, Performance T. Supporting electricity sector reform in Libya. 2017;(07).

[7] Turkish G, April S, Accord N, Army LN, Turkish G. Tripoli' s Electricity Crisis and its Politicisation Policy Brief. 2020;(April).

[8] Data sheet of Al-Zawia combined cycle power plant 2019.

[9] Data sheet of Western Mountain power plant 2019-2020.

Chapter 5

Smart Monitoring of Flat Wheel in Railway Using Optical Sensors

Preeta Sharan, Manpreet Singh Manna and Inderpreet Kaur

Abstract

The need for improved safety, reliability and efficiency is one of the most important aspects of the railway industry worldwide. Optical sensors can be used in smart condition monitoring system that can allow real time and continuous monitoring of the structural and operational conditions of trains. Railway monitoring is carried by the use of Fiber Bragg Grating sensors which measures strain, vibration, temperature, acceleration in continuous manner. This chapter covers introduction and working of optical sensors, Finite Element Analysis of rail-wheel geometry and health monitoring of rail wheel. FBG as optical sensor is well known for its advantages such as easy multiplexing, wavelength encoding and multiparameter sensing, immune to electromagnetic interference, reliability, flexibility. Sensitivity of optical sensor in compare to traditional sensors goes as 1.2 pm/με and 10 pm/με for strain and temperature sensor at 1550 nm of wavelength.

Keywords: Rail Wheel Geometry, Wheel Flat, Strain Analysis, Fiber Bragg Grating Sensors

1. Introduction

An optical sensor converts light rays into electronic signals. It can be used as a smart condition monitoring system in different fields such as civil, mechanical, electronics, biomedical for real time and continuous monitoring. Many parameters of railways uch as weighing of wagons, constant speed and direction of the train can be tracked and monitored in continuous manner with the help of optical sensor. Health monitoring of rail and wheel is another important aspect which is covered in this chapter. Railway monitoring is carried by the use of Fiber Bragg Grating sensors which measures strain, vibration, temperature, acceleration in continuous manner [1–5]. Different sections of chapter covers introduction and working of optical sensors, application of FBG in railways, study of rail-wheel geometry by various methods and last is defects in wheel and how optical sensor detects it.

2. Fiber Bragg Grating

A Fiber Bragg Grating sensor is a distributed Bragg reflector, i.e. a periodical variation of refractive index, inside the core of optical fiber, able to reflect a particular wavelength of light and transmit all the others. WhenFBG is subjected to external factorssuch as pressure, vibration, temperature, stress and strain, refractive index and grating period varies, there will be corresponding changes in the reflected wavelength. Since the parameter of measurement is the wavelength of light which is not affected by electromagnetic fields, the process is immune to electromagnetic interference and hence is intrinsically more stable than any electrical monitoring system as explained in [6]. The reflected wavelength can be calculated as

$$\lambda_B = 2n_{eff} \wedge \quad (1)$$

Here λ_B is Bragg's Wavelength, is effective refractive index and Λ is periodic variation of FBG. **Figure 1** shows the general schematic diagram of Grating sensor and spectral response of the Fiber.

Bragg Gratings can be written into single mode fiber with inner core diameter 5 to 9 μm and cladding diameter of 125 μm. Core is made up of silicon doped with germanium whereas cladding is pure glass material. Due to this there is high difference in refractive indexes between inner core and cladding thereby making light to propagate inside the inner core only. In [7] fabrication method is given using Holographic method and Phase Mask Method. Phase mask method is commercially used to fabricate optical fiber as in holographic method more stable setup is required with good coherence light source. In Phase mask technique fiber is exposed to a pair of interfering UV beam then there are regions of constructive interference and destructive interference, the first region corresponds to high UV intensity and refractive index will increase whereas in destructive interference intensity of UV light is negligible, there is no index change. This exposure to an interference pattern will result in a periodic modulation of refractive index along the core of the fiber and gratings are formed which reflects a particular wavelength of light and transmits all others. The reflected wavelength is known as Bragg's Wavelength.

2.1 FBG as optical sensor

Fiber Bragg Grating can be used as various sensors based on the fact that Bragg Wavelength changes with the change in refractive index or period of the grating as given in [8]. Here it is explained how Bragg grating is applicable as strain and temperature sensor, pressure sensor and stress sensor. When FBG is subjected to various external parameters such as pressure, strain, temperature, displacement, load and vibration there is a change in the period of grating, either elongates or compressed and effective refractive index also varies, due to this there is a shift in Bragg wavelength. **Figure 2** describes the reflectivity response, explained in [8].

FBG can measure strain and temperature by means of detecting changes in the reflected wavelength of light which can be calculated as given in Eq. (2).

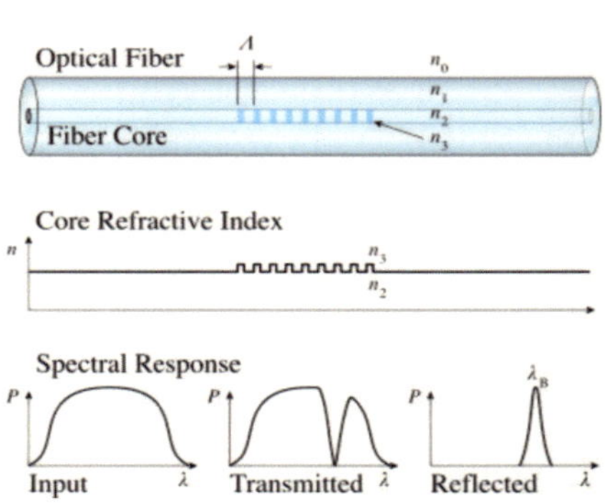

Figure 1.
Fiber Bragg Grating structure within the core of Optical Fiber.

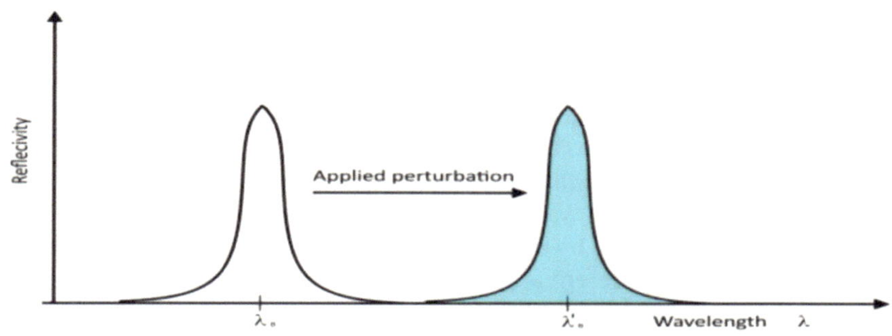

Figure 2.
New Reflectivity Response of FBG with the change in wavelength.

$$\Delta\lambda B = \lambda B(1 - \text{Pe})\,\Delta\varepsilon + (\alpha + \zeta)\,\Delta\text{T} \tag{2}$$

Where Pe is photo elastic coefficient of the fiber, α and ζ are thermal expansion and thermo optic coefficient of the fiber material, $\Delta\lambda B$ is new wavelength. At 1550 nm centre wavelength, the wavelength strain and wavelength temperature sensitivities are 1.2 pm/με and 13 pm/°C. For measuring axial strain along the fiber due to applied pressure P is given in Eq. (3)

$$\varepsilon = \frac{\text{P}(1 - 2\upsilon)}{\text{E}} \tag{3}$$

here, ε is strain, P is pressure, v is Poison Ratio and E is Young's Modulus. The intrinsic pressure sensitivity of a bare FBG is only 3.04 pm/MPa, which is too low for the practical pressure measurement, methods proposed to enhance the pressure measurement sensitivity indirectly, such as embedding FBG in polymer, soldering metal-coated FBGs on a free elastic cylinder, and attaching the FBG fiber to a diaphragm given in [9].

2.2 Fiber Bragg Grating in railways

FBG can be used for health monitoring in railways as it can monitor different train parameters such as speed of the train, wagon weight, axle count and determine the rail-wheel condition and bogie health monitoring as given in [1, 2, 10]. Monitoring these parameters in railway continuously at minimum expenditure can help us to build a SMART RAILWAY SYSTEM for the betterment of mankind. This is possible by the use of Fiber Bragg Grating sensor which when compared with other electrical sensors such as strain gauges or accelerometers, optical fiber sensor has many advantages such as easy to install, more durability and reliability, cost effective, has multiplexing and de multiplexing characteristic and above all it is immune to electromagnetic interference.

In [3] how FBG act as novel optical sensor to detect flat wheel and weigh in motion is explained. Field trials have been carried out along the rail. FBG sensor clamped to rail, detects the vertical forces generated by the wheel rail contact in terms of wavelength shift in FBG. This shift gives a lot of information about the train in transit, such as wheels weight and their defective status in real time scenario. FBG as strain sensors gives the wavelength shift, characterized by a sequence of pulse. It is observed that pulse related to the wheels of engine gives large wavelength shift than the pulses of empty wagon. FBG for train axle count is described in [4]. Two parameters are considered here train detection and train control. Track circuits are used for train detection, by means of simple open and close circuit principles. FBG as optical fiber sensor detects number of axles when train passes over the rail. Sensor is installed on rail to measure the strain change in the rail upon the passage of trains. It is also given that a conspicuous and distinctive peak can be identified in the output strain signals measured by the grating sensor. In [5] optical Bragg Grating Sensor measures the ultrasonic guided waves in subway rail sample. Here FBG sensor detects the ultrasonic waves generated by the ultrasonic actuator placed along the rail, to detect cracks in rail. Two different approaches are used FBG sensor and piezo electric transducer (PZT) transducer to capture ultrasonic waves. The actuator was excited by 40 KHz Gaussian sine type with 13 cycles, 200Vpp amplitude. In compare to common PZT transducer, advantage of the use of FBG sensor is that it can be located far away from ultrasonic sensor as it can capture the waves transmitted at light speed without electromagnetic interference. It is concluded in this work that FBG sensors are capable to measure ultrasonic guided waves in rail transport monitoring.

3. Rail-wheel analysis

The railway train comprises of engine and number of wagons. The engine has 12 wheels, 6 axles. The wagon has eight wheels, 4 axles. This is the description of Indian Railways. Below **Figure 3** explains the rail cross section. Main parts of rail are

1. Upper region known as head of the rail, wheels of the wagon rotates on it.

2. Middle part between the head and foot is web region, mostly sensor is attached to this part.

3. Lower part is foot part

Rail track is an important part of the railways which gives persistent and level surface for movement of the train made up of steel. Rail act as guide for the wheels in a lateral manner. When wheels of the wagon rotates on the rail there are certain forces exerted on the surface of the rail and different static and dynamic stress and strain are developed on the rail. The stress distribution is an important factor at the rolling interfaces between rail-wheel which depends on the geometry of the contacting surfaces, loading and boundary conditions and material properties of wheel and rail as given in [11].

Defects on the rail surface occur due to the cracks generated by repeated passage of trains and may increase with time. When wheel rotates on the rail track the contact between the wheel and the rail, many problems may arise which are

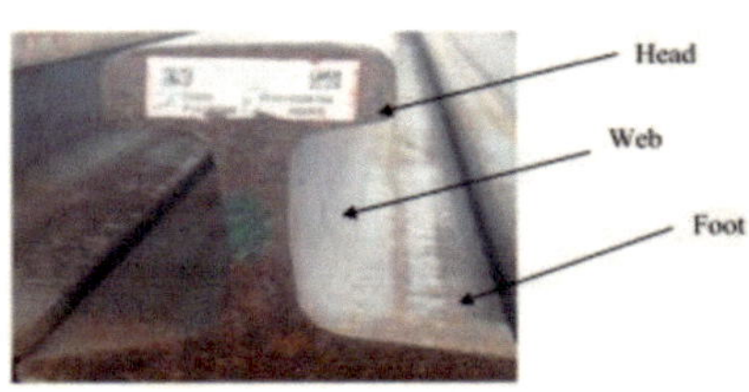

Figure 3.
Rail-wheel model.

1. Flat spot on the wheel

2. Cracks on the rail

3. Thermo-elastic plastic behaviour in contact

The wheel and rail interaction can be carried out by different methods

a. Hertz contact theory

a. FEA Analysis using software

3.1 Hertz theory

Railway wagon is a complex mechanical structure with body, wheel and axle as three main parts. It has several degree of freedom which includes linear and non-linear springs and various types of damping system. When wheel rotates on the rail, different forces acting on the rail surface are vertical forces, horizontal longitudinal and transverse forces as given in [12].

Figure 4 gives the various forces acting on the railway track. The static force acting on the mass is the weight of bogie and car $(m_b + m_c)$gand Hertzian spring acting in the wheel/rail contact area is kHz. The 1/4 bogie model is a two-mass model, in which the other mass (mass of wagon m_b) is added over the mass in the 1/2 axle model that simulates the structure of the bogie, between the two masses is located the vehicle's primary suspension (k1, c1), and the static force acting on the superior mass is the weight of the car mcg.

Figure 5 describes the vehicle model as given in [13]. The 1/8 vehicle model is a three-mass model, in which the other mass (mass of wagon m_c) is added over the masses of the 1/4 bogie model that simulates the box of the vehicle, the vehicle's secondary suspension (k2, c2) is located between the upper masses.

3.2 FEA analysis using ANSYS software

Ansys 15.0 is a mechanical tool to perform finite element analysis (FEA) for structural analysis. Different types of model can be designed and analysed using this software [14]. The ANSYS software can be used for Finite Element Analysis of rail-wheel part. It has three working sub parts. They are

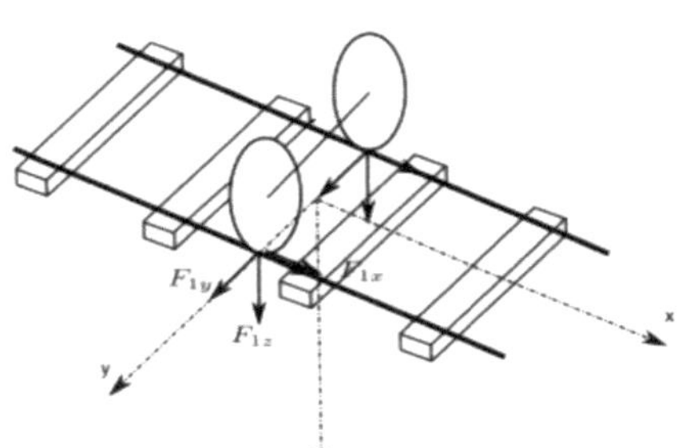

Figure 4.
Different forces applied on the rail.

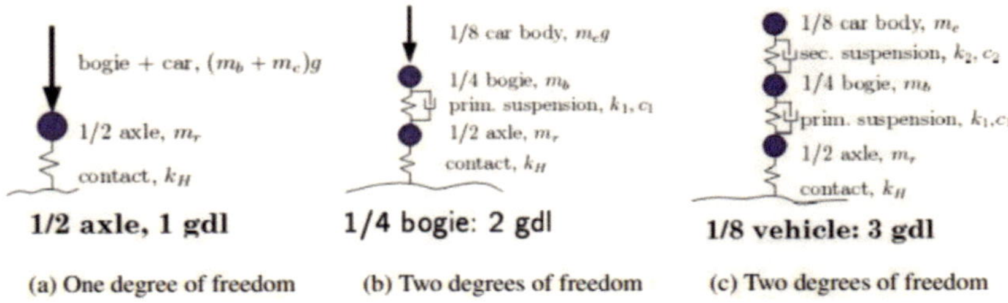

Figure 5.
Vehicle model.

A. In Preprocessor

i. Material properties are defined like structural steel, ballast, concrete.

ii. Element model is assigned.

iii. Build analysis model

B. In Processor

i. Boundary conditions are defined with the application of loads.

ii. Define the load characteristics.

iii. Control the convergence mode.

iv. Input parameters are taken here like pressure, bearing load, rotational velocity etc.

v. Start evaluation

C. Postprocessor

i. Evaluate the results.

ii. Draw diagram with results.

iii. List the results, note the values.

Mesh analysis of rail wheel model is given in **Figure 6**. Friction coefficient between the wheel and rail is 0.3. After defining material properties, the model

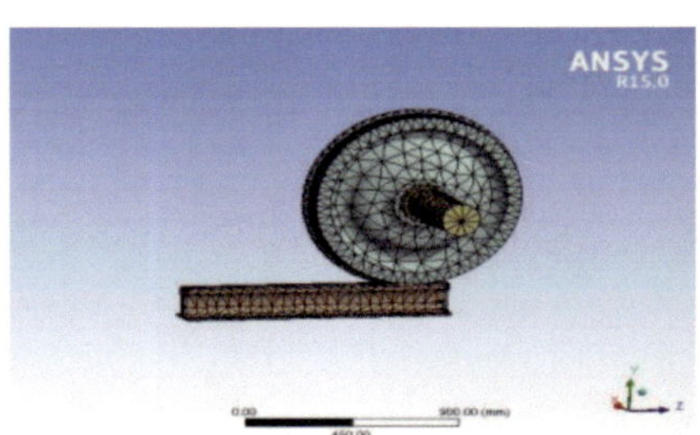

Figure 6.
Mesh analysis of rail and wheel.

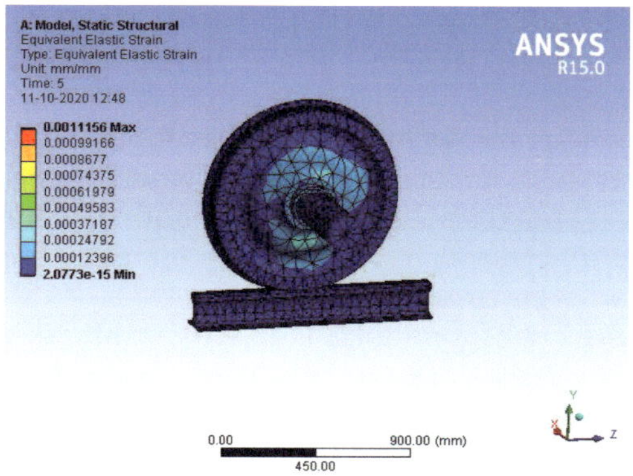

Figure 7.
Equivalent elastic strain.

is meshed. To obtain accurate results near the rolling contact surface, rail-wheel contact areaa fine mesh is done with element size 50 mm. **Figure 7** shows the output result of equivalent elastic strain.

To simulate 3D analysis of single rail wheel contact, following steps should be considered

1. Accurate modelling of railway parts wheel rail and sleepers
2. Material modelling, elastic or elastic–plastic, uniform and elastic
3. It is nonlinear and temperature independent
4. Boundary and loading conditions
5. Combinational, rotational and linear motion of wheel
6. Defining region of contacts to impacted elements
7. Considering straight track and standard rail profile
8. Defining nodal force and output parameters as equivalent elastic strain and stress with total deformation

4. Problems in rail-wheel: flat wheel

A wheel flat is a fault in the rail wheel shape, a geometric chord developed on the wheel surface caused when wheel slides on the rail or when wheel or axle has stopped rotating. There are many reasons for flat spot on the wheel which may include use of emergency brake or slip and slide conditions. Due to this wheel gets locked while train is in motion. Faulty brakes or wheel set bearings is another reason for flat spot. Wheel flat is also known as spalling or shelling or out of roundness. Flat spots on the wheel of the wagon is a primary cause of track and wheel quality deterioration as given in [15]. This is a very common problem which railway industry faces and leads to accidents. When a defect wheel rotates on the rail, large dynamic forces are produced, this increases acceleration levels on the track and vibrations are produced which is transmitted to the rolling stock. Force produce by flat wheel can damage the rail, suspension system, frame and also the body of rolling stock as given in [16]. If we look deeper into the mechanism of wheel flat loading, when the wheel with flat defect rotates on the rail, there is no contact between the wheel and the rail and it rises above the rail for a very short duration but recovery of this contact results in a high impact force which can damage the rail surface. **Figure 8** gives the mechanism of how flat wheel affects the rail. The vertical force of these wheels are ten times higher than the normal wheel as explained in [17]. Velocity of the train and weight of the wagon increases the dynamic forces produced by the flat wheel increases.

To detect wheel with out of roundness is an important task which have been carried out from past many years by researchers using various methods and different sensors. In [18] experimental analysis has been carried out in Lithuanian railways lines to measure the vertical force impact between the rail and the wheel. There are two types of sensors used, strain gauges to measure force and accelerometers to measure rail motion. Signal processors analyse the data to separate the wheel with flat irregularities. Here strain gauges act as wheel sensors that weigh each wheel of the train as it passes by. It has been observed that duration of the highest momentary force action in the wheel and rail contact depending on the velocity of the train is for few milliseconds and even less. This has been fixed by Wheel Impact Load Detector (WILD). Another method to detect wheel with flats is by using Fiber Bragg Grating optical sensor. As discussed earlier FBG sensors are widely used for their several advantages when compare to electrical sensors such as flexibility, durability, long life time and most important immune to electromagnetic interference and multiplexing property, many FBG with different grating on a single optical fiber. **Figure 9** shows the flats on the wheels restricted to 60 mm and depth 0.9–1.4 mm. In railways optical sensor can be easily installed along the track for various kilometres to detect several parameters of train. FBG readings is taken by the interrogator further connected to the computer.

In [19] Fiber Bragg Grating strain sensor has been used to study the vibrations from the passing train. **Figure 10** shows the reading of the sensor in terms of micro strain and time which is installed on the track with a passage of train with 12 wagons. Each wagon consist of 8 wheels, 4 axles, hence first four peaks on the left side correspond to first wagon. Here it is observed that wagon 6 and 8 have noisy or abnormal vibrations, hence the wheels of these wagon may have defects (**Figure 10**).

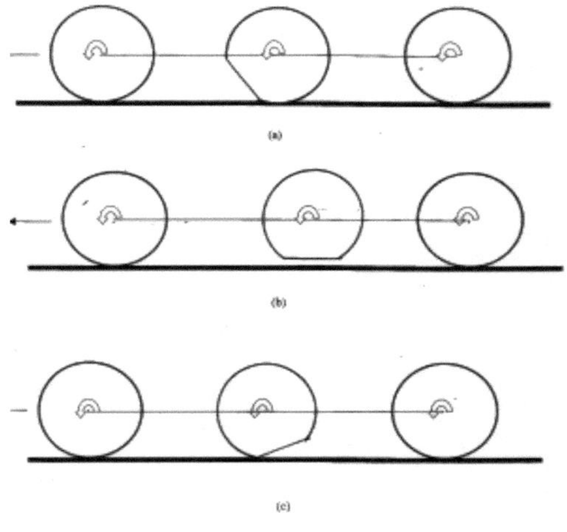

Figure 8.
Lift and hit motion of the wheel. (a) Rotation of flat wheel (b) centre wheel lift up (c) hit motion of wheel.

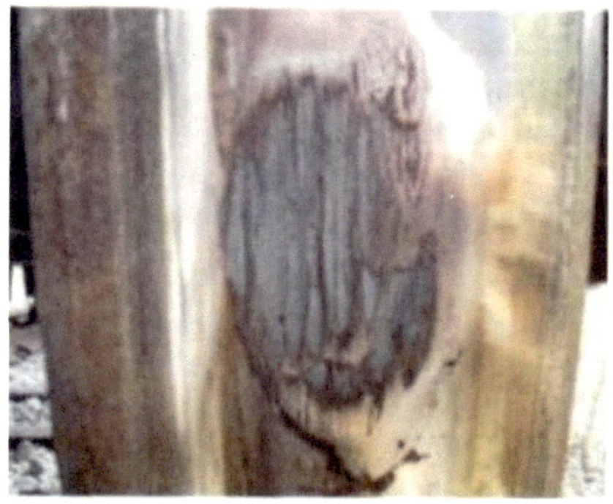

Figure 9.
Flat portion on the wheel.

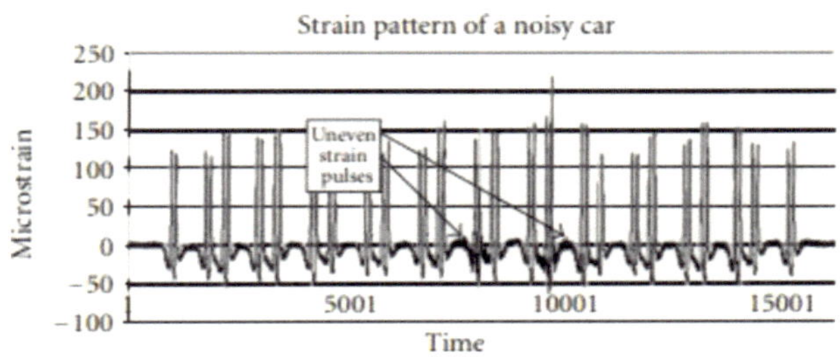

Figure 10.
Strain measured by Fiber Bragg Grating [19].

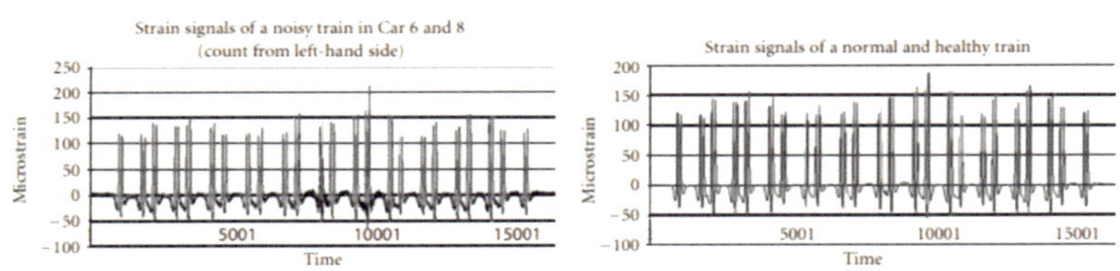

Figure 11.
Strain signal from noisy train and normal train [19].

Strain measurement at the track shows that the noisy trains may have imperfect wheels. It can be noted that defect wheel will produce an uneven strain impulse on the track as shown in **Figure 11**. Good wheel strain signal will be even and periodic in nature.

Fiber Bragg Grating as strain sensor is an effective means to distinguish between healthy and unhealthy wheels. It measures the strain in terms of shift in Bragg's wavelength and detects the vibrations from noisy trains which can be analysed further to find wheels with flat spot.

5. Conclusion

In this chapter, the basic rail wheel model is given and how we can analyse using various mechanical tool is explained, output includes elastic stress and strain and total deformation produced between the rail and contact. Various parameters in railway is considered and problems related to it are discussed in this chapter. The readings of Fiber Bragg Grating sensor in flat wheel can be further used for signal processing by using filters such as Low Pass Filter, High Pass filter to remove noise in the output and data acquisition by which we can study various parameters of wheel and rail and can also monitor the rail wheel condition. It can be concluded that in real time various dynamic parameters of train like speed, gross train weight, axle weights, axle spacing can be determined by strain monitoring data which optical sensor Fiber Bragg Grating can read.

Author details

Preeta Sharan[1], Manpreet Singh Manna[2] and Inderpreet Kaur[3]*

1 Department of ECE, The Oxford College of Engineering, Bangalore, India

2 Department of EIE, SLIET, Longowal, Punjab, India

3 Department of EE, GNDEC, Ludhiana, Punjab, India

*Address all correspondence to: inder_preet74@ieee.org

References

[1] M.L. Filograno et al. Real time monitoring of railway using Fiber Bragg Grating sensors. Proceedings of the joint rail conference. 2010.

[2] Cristian Vendittozzi et al. Static and dynamic weighing of rolling stocks by mean of a customized FBG- sensorized patch. International Journal of safety and Security Engineering. February 2020; 83-88.

[3] Antonio Iele et al. Fiber Optic Sensing System for Weighing In Motion (WIM) and Wheel Flat Detection (WFD) In Railways Assets: The TWBCS System. 8th European Workshop on Structural Health Monitoring (EWSHM 2016). July 2016.

[4] Chu-liang Wei et al. A Fiber Bragg Grating Sensor System for Train Axle Counting. IEEE Sensors Journal. 2010; 1905-1912.

[5] Wagner Francisco Rezende Cano et al. Evaluation of FBG sensors to measure ultrasonic guided waves in rail transport monitoring. IEEE Xplore Proceedings. 2017.

[6] Myra Lydon et al. Improved axle detection for bridge weigh-in-motion systems using fiber optic sensors. J civil structural health monitoring. July 2017; 325-332

[7] Andrija Milojevic et al. Application of FBG Sensors in Smart Railways. XV International Scientific Expert Conference on Railways. XV International Scientific Expert Conference on Railways. October 2012.

[8] Carlo Edoardo Campanella et al. Fibre Bragg Grating Based Strain Sensors: Review of Technology and Applications. September 2018.

[9] Jun Huang et al. A Fiber Bragg Grating Pressure Sensor and its Application to pipeline leakage detection. 2013.

[10] Radek Martinek et al. Fiber Optic Bragg Sensors for the Rail Applications. International Journal of Mechanical Engineering and Robotics Research. May 2018.

[11] Roop Lal et al. Stress analysis at contact region of rail-wheel: review. Vth International Symposium on Fusion of Science and Technology. January 2016.

[12] NS Vyas, AK Gupta. Modelling Rail Wheel-Flat Dynamics. WCEAM 2006. Paper 233

[13] J.M. Goicolea et al. Dynamic Analysis of High Speed Railway Traffic Loads on Ballast and Slab Tracks. Proceedings of the Tenth International Conference on Computational Structures Technology. 2010.

[14] B. Jagadeep et al. Stress Analysis on Rail Wheel Contact. International Journal of Research in Engineering, Science and Management. May 2018; 47-52

[15] Roveri et al. Remote Condition Monitoring of Railway Track using FBG Sensors. Proceedings of ISMA 2014; 3527-3541.

[16] Massimo Leonardo Filograno et al. Wheel flat detection in high-speed railway systems using Fiber Bragg Gratings. IEEE Sensors journal. December 2013; 4808-4816

[17] R. Kasanna et al. Design and Analysis of Indian Wheel-Rail Assembly for Super Elevation, International Journal for Research in Applied Science and Engineering Technology. July 2017.1407-1418.

[18] Olegas Lunys et al. Analysis of freight wagon wheel failure detection in Lithuanian Railways. 9th International

Scientific Conference Transbaltica 2015; 64-71

[19] C.C. Lai et al. Development of Fiber Optic Sensing System for Train Vibration and Train Weight Measurements in Hong Kong. Journal of Sensors. Vol. 2012.

Chapter 6

Evaluation of the Level of Prognostic Competence with Fuzzy Logic Systems

Juriy Andreevich Tashkinov

Abstract

The chapter deals with main components are part of the prognostic competence. The chapter discusses possibility of using intelligent systems to model the predictive competence of a civil engineer. The analysis of the study of the problem was carried out using content analysis A model of the development of prognostic competence of a civil engineer using the Matlab software package with the Fuzzy Logic Toolbox module has been developed. A step-by-step forecasting technology, as well as possible errors in creating a forecast by an expert and ways of solving them, are proposed. In conclusion the chapter says most of all the level of formation of predictive competence depends on activity component.

Keywords: civil engineer, prognostic competence, pedagogical forecasting, intelligent systems, Educational Data Mining

1. Introduction

There are only the last two major catastrophes, which allegedly occurred due to the fault of civil engineers and other specialists from related industries: fire in the mall "Zimniya Vishnia" on March 25, 2018 on Kemerovo and the collapse of the Morandi automobile bridge on August 14, 2018. But this list could be continued with a huge number of examples. That's why the formation of prognostic competence of construction university students as future civil engineers is very important. Predicting a future catastrophe is easier than eliminating its consequences. Every specialist, especially representatives of engineering professions, must possess prognostic skills [1]. Pedagogical forecasting, in contrast to other branches of science about the future state of the object under study, is fraught with a number of difficulties. This is due to the complexity of the predicted object: the changeable personality of the student, the development of which depends on a number of factors, both external and internal. Formalization simplifies the object and does not allow to include all indicators, factors and relationships in the model.

2. Analysis of the problem field

To prove the relevance of the chosen direction of research, the analysis of publication activity for a given keyword "pedagogical forecasting" in the world literature

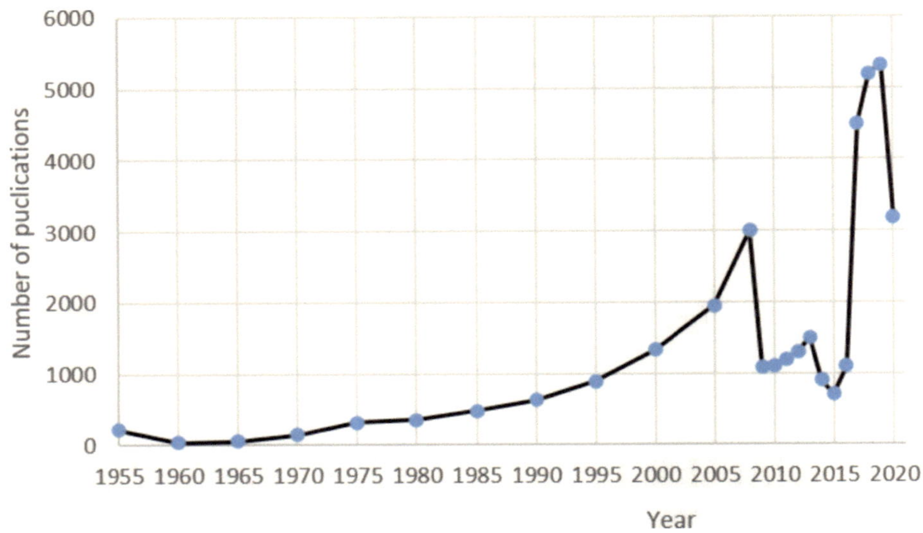

Figure 1.
The number of publications on the practice of using pedagogical forecasting (according to the Google Scholar database).

was carried out; found the number of articles published in a certain period of time, the result is plotted in **Figure 1**.

We see the growing interest of scientists in the use of pedagogical forecasting tools in 2008, 2013 and 2019, which is associated with technological breakthroughs in the field of computer technology. The introduction of computer technology in the educational system is a priority of modern public policy. Intelligent systems are widely used. This technology is application promising in the field of pedagogical forecasting and the formation of the prognostic competence of a specialist. The theory of forecasting is considered in the works of I.V. Bestuzheva-Lada et al. [2], K. Beecher, D. Berry, J. Maliekal, O. Yasar, R.S.J.D. Baker [3], A. Dutt [4], B. Oancea, C. Romero, S. Ventura, M. Pechenizkiy et al. [5] are engaged in research in the application of Educational Data Mining technology for modeling educational processes. However, only a small amount of work is devoted to the formation of the prognostic competencies of a civil engineer.

3. Task

To build models of the civil engineer predictive competence formation with using the Matlab R2014a software package with the Fuzzy Logic Toolbox module depending on different factors and components.

4. Development of methodology

4.1 Prognostic competence

Forecasting is a cognitive activity aimed at revealing the features and characteristics of the personality development processes of a student of a construction university and the consequences expected from them; an indication of the path and conditions for the implementation of foresight. Predictive competence [6] is the result of education including the level of professional preparedness of a student, knowledge, skills, forecasting skills allow to possess the knowledge and skills of the forecasting process, selection and logical processing of necessary information, analysis and determination of trends have the skills of planning, goal setting, programming, design; develop the ability to understand opportunities, abilities, ways to improve.

4.2 Component composition of professional competencies

Table 1 shows the characteristics of each of the components of the development of competence [7].

A-level of prognostic competence and its components will be estimated at 100 points, B-level – at 75 points, C-level – 60 points, formal level below 60 points (F, we estimate in 0 points).

Analysis of the scientific literature on the relationship of knowledge, skills and personality traits in the overall effective assessment of training allowed us to identify the following weights of the components of competence: the cognitive component is.2; activity component is.5; motivational component is.15, communicative component is.15 [8].

4.3 Rules and methodology of modeling of dependence of prognostic competence general level depend on the development of its components

We will use the methodology is described in [9] and theoretical principles is described in [10].

Matlab R2014a software package with Fuzzy Logic Toolbox module will be the main tool of our research. We will use the Mamdani algorithm mode. We will need

Levels	Component composition of professional competencies			
	Cognitive	**Activity**	**Motivational**	**Communicative**
C	the student is ready to reproduce the studied algorithms for solving professional problems	the student is ready to solve professional problems according to a given algorithm	the student is ready to modify the existing sample of tasks for a given algorithm	the student is ready to adequately analyze and evaluate their actions and the actions of members of the educational team
B	the student is ready to offer similar algorithms studied in solving standard problems	the student is ready to solve standard professional problems by analogy with the studied algorithm	the student is ready to upgrade professional tasks according to original algorithms	the student is ready to plan, organize and adjust their activities and the activities of the members of the educational team
A	the student is ready to develop his original algorithms for solving professional problems	he student is ready to apply original algorithms in the course of professional activity	the student is ready to simulate non-standard situations containing functional professional novelty	the student is ready to be responsible for the results of professional activity, to model and predict their further professional development

Table 1.
Characteristics of the levels of development of professional competencies of future engineers.

4 input variables to describe the level of development of the prognostic competence component and 1 outgoing variable to describe the general level. The type of the membership function is gaussmf, since most pedagogical phenomena are subject to the normal Gauss distribution [10].

To formulate the rules of fuzzy logic we will use the MS Excel 2007. We have formulated 256 fuzzy logic rules (4^4), but in article we will present only a few examples:

- If Cognitive component is in B-level AND Activity component is in A-level AND Motivational component is in B-level AND Communicative component is in B-level THEN General level of prognostic competence formation is in B-level;

- If Cognitive component is in B-level AND Activity component is in B-level AND Motivational component is in A-level AND Communicative component is in formal level THEN General level of predictive competence formation is in C-level;

- If Cognitive component is in A-level AND Activity component is in B-level AND Motivational component is in B-level AND Communicative component is in C-level THEN General level of prognostic competency formation is in C-level;

- If Cognitive component is in A-level AND Activity component is in C-level l AND Motivational component is in B-level AND Communicative component is in formal level THEN General level of prognostic competence development is in C-level.

4.4 The most factors influent on prognostic competence

The authors [11] concluded that pedagogical factors influence the formation of prognostic competence by 31%, psychological group of factors – by 28%, social and economic conditions: 33%.

4.5 Rules and methodology of modeling of dependence of prognostic competence general level depend on the development of the most factors

We used the Mamdani algorithm mode. We needed 3 input variables, each of which corresponded to a factor influencing the formation of predictive competence, and 1 outgoing – the level of competence formation. The type of membership function for predictive competence is gaussmf, since most pedagogical phenomena are subject to the normal Gauss distribution [10]. For factors that determine the development of prognostic competence – the type of membership function – trimf.

The evaluation interval for the influence of each of the factors is 0–100, which corresponds to conditional estimates: "low level" (F), "below average" (D), "average level of efficiency" (C), "above average" (B), "High level" (A). The creative level of the development of predictive competence will be estimated at 100 points (A), productive – 75 points (B), reproductive – 60 points (D), formal level – below 60 points, but above 20 (F).

To formulate the rules of fuzzy logic we will use the MS Excel 2007. We have formulated 125 fuzzy logic rules (5^3), but in article we will present only a few examples:

	Stages	Organizational and pedagogical conditions	Possible sources of forecast errors	Ways to troubleshoot errors	Application time
	I. *Flashback stage*				
1	preliminary stage: obtaining the task and forecasting constraints; choice of subject and object	study of state standards and other regulatory documents; Web Mining; Text mining	the customer's idea of the predicted object can be vague, which leads to an unclear formulation of the task	are eliminated by an expert at the next stage	before the start of the educational process
2	predictive justification: the stage of setting goals and objectives of pedagogical forecasting; tracking the results of preparatory information; formation of expert groups and assessment of the competence of expert analysts	training experts in forecasting and methods of computer pedagogy, specifics of forecasting in the professional training of students in construction areas of training; Web Mining; Text mining	research of a new object for an expert can lead to incorrect formulation of goals and objectives, which can lead to unexpected results (or their absence); incorrect understanding by the expert of the tasks set by the customer of the forecast	clarification of goals after conducting pre-predictive orientation; discussion of goals and objectives with the customer	before the start of the educational process
3	pre-predictive orientation stage; tracking the results of preparatory information; formation of expert groups and assessment of the competence of expert analysts	conducting a test to determine the level of formation of an expert's predictive skills; development of a fund of predictive and evaluation tools; non-formalized methods, clustering, Visual Mining ("Chernov's faces"), the Delphi method (to assess the competence of an expert); brainstorming; Organizational EDM; maps of the future and SWOT analysis; informal methods	the application of the algorithm by different teachers can lead to different results; the existence of two or more interpretations (sometimes contradicting each other) of the same phenomenon; abrupt changes in the construction industry and in the training trends of future civil engineers (irregular mistakes)	development of a standard procedure for conducting pre-forecast orientation; expert assessment of the research object; exploring different approaches; study of new literature at all stages of forecasting	after the introductory campaign; after 1 session; upon completion of 1 year of study
4	the stage of systematization of the collected information about the educational process in a construction university; analysis of constraints when creating a forecast	building maps of the future by students, teachers and representatives of construction companies; non-formalized methods, clustering, Visual Mining ("Chernov's faces"), the Delphi method (to assess the competence of an expert); brainstorm	the most complete list of characteristics, covering all manifestations without exception, practically cannot be realized	Highlighting important characteristics	after the introductory campaign

Stages		Organizational and pedagogical conditions	Possible sources of forecast errors	Ways to troubleshoot errors	Application time
5	stage of development of the research program; study of possible risks	precise formulation of organizational and pedagogical conditions, appointment of those responsible for implementation and deadlines; Organizational EDM; maps of the future and SWOT analysis	teaching work is carried out in conditions of incomplete information, which are difficult to predict and take into account. Sources of uncertainty can be random factors, the mood of students and the teacher, sometimes these are external factors; pedagogical processes are probabilistic in nature, which causes difficulties in forecasting	standardization of procedures for formalizing pedagogical phenomena	after 1 session
	II. *Stage of analysis of diagnostic information*				
1	the stage of compiling a list of model indicators based on data from various sources	questioning of teachers, students, representatives of construction companies in order to identify the main external factors; study of literary sources; non-formalized methods; “The logical square of the competences of the future civil engineer”; “Competency profile”; descriptive analysis; classification; Visual Mining; fuzzy logic systems	forecasting OR requires an analysis of many factors, conditions and circumstances; what you are looking for is not clearly formulated. direction of training 08. 03. 01 “Construction” includes a lot of specializations. Students from different groups study different special courses, which complicates the collection of statistical information. Differences in predicting the RR of master’s and bachelor’s students; full-time and part-time students	reducing the number of factors, creating object restrictions; creating different models for each training profile	2 course

DOI: http://dx.doi.org/10.5772/intechopen.99919

Stages		Organizational and pedagogical conditions	Possible sources of forecast errors	Ways to troubleshoot errors	Application time
2	stage of collecting forecast background data (economic; sociological; socio-cultural; political and international; legal; pedagogical; demographic; natural; scientific and technical; organizational)	maximum digitalization of information: statistics on student attendance at classes, intermediate control data in all disciplines, information on extracurricular activities (for example, volunteer); maintaining a portfolio of a student and an academic group; questionnaire survey of teachers, students, representatives of construction companies in order to identify the main external factors; study of literary sources about the forecast background for each region; postulation using literary sources; system analysis; Real - Time EDM; Visual DM; factor analysis; "Forecast background profile"; fuzzy logic systems	in the literature, indicators of the forecast background for a civil engineering university have not been sufficiently studied; underestimation by an expert of the degree of background influence; different background indicators affect different students differently; the indicators of the forecast background for students of different universities differ; indicators of the forecast background are difficult to formalize	study of economic, psychological and other non-educational literature. strict adherence to forecasting technology procedures; studying the personal affairs of students, obtaining information from the curators of the groups, creating an individual forecasting background; cluster division of students into conditional groups experiencing the maximum influence of the same types of forecast background	2 course
3	The stage of taking into account factors that negatively affect the achievement of educational results	Educational conversations with students who have academic arrears	"Survivor 's Bug"	About asking expelled students about the reasons that may have influenced their failure to achieve educational results	
4	the stage of generalizing the preliminary list of indicators to a type that will help to present the best type of pedagogical forecast	filling out the "form for assessing the key competencies of students in construction areas of training" (see Table B. 3, Appendix B); building a "Competency Calculator"; Real - Time EDM; Visual DM; OLAP; individual IP competency profile	educational outcomes for full-time and distance learning differ	correction of the forecast when students switch to distance learning	2 course

Stages		Organizational and pedagogical conditions	Possible sources of forecast errors	Ways to troubleshoot errors	Application time
5	diagnostic information analysis stage	self-analysis of the formation of prognostic competencies; conducting a test to determine the level of formation of students' prognostic skills; solving prognostic problems; Delphi method; dividing students into clusters (conditional groups); Real - Time EDM; Visual DM; OLAP; dynamic individual IP competency profile; clustering	it is impossible to prove the consistency of a formal system by means of this system itself (Kurt Gödel). in a real-life problem, none of the functions is known for sure - approximate or expected values are given. in the course of forecasting OR, it is necessary to take into account both the uncertainty of the data, the system, and simulate the potential contributions to the predictive uncertainty; in the case of long-term forecasts, it is impossible to reliably state when the system will cease to function; Arrow's paradox about the impossibility of an absolutely objective assessment by equivalent experts	Application of means of computer pedagogy; In order to get rid of the uncertainty, it is necessary to fix the functions, while losing the accuracy of the description of the problem. carry out operational dynamic tracking of the state of the predicted object; m Selecting a person responsible for forecasting ("dictator")	after 2 course
	III. *Stage of model building*				
1	stage of building a model of educational outcomes of future civil engineers and selection, suitable forecasting methods	filling in the "prognostic portfolio of students of construction areas of training" individual for each student pedagogical SWOT analysis; multiple regression; creation of a system of equations; SWOT analysis	statistical methods are well developed for one-dimensional random variables. in multivariate statistics, for lack of something better, poorly grounded heuristic methods are often used; if there is a simple dependence, then its form is not known in advance.	comparison of results obtained in different models or by several experts	before the winter session in the third year
2	Stage of completion of the basic model				

Stages		Organizational and pedagogical conditions	Possible sources of forecast errors	Ways to troubleshoot errors	Application time
	IV. *Prospect stage*				
1	the stage of constructing time series of forecasting for each indicator of the base model (extrapolation)	an expert must adhere to the basic principles of pedagogical forecasting: adequacy, alternativeness, variability, verifiability, continuity, consistency, consistency, efficiency; checking the periods of foundation and lead time; construction of time series; Real - Time EDM; OLAP; Visual EDM	You can meet the "Oedipus effect" of forecasting: people tend to believe in predictions, and strive to fulfill the created forecast at any cost (the so-called "self-fulfilling forecast")	We do not recommend without unnecessary need to inform students about low predicted educational results, so as not to underestimate their stress resistance, but positive predictions can be voiced, since this can increase the motivational component of competencies.	Before the start of the 4th course
2	the stage of determining the absolute optimum with conditional abstraction from the limitation of the forecast background				before the winter session in the fourth year
3	stage of pedagogical forecast verification and examination; analysis of sources of forecast errors; assessment of the consequences of applying the forecast and compliance with all forecasting principles	verification of the reliability, accuracy and forecast error, according to the real level of competence formation in a bachelor's degree graduate; dichotomy (including multidimensional); expert assessments; statistical methods; Visual EDM	"The illusion of predictability: several experts come to the same wrong conclusion, because the mathematical model does not fully reveal the essence of the phenomenon under study, due to uncertainty, the emergence of the dependent variable;	make a "blind" forecast: experts should not be familiar with the prediction of colleagues	After the students graduate from the experimental groups of bachelor's degree
4	forecasting model adjustment stage	re-evaluation of expert forecasting knowledge; clarification of the organizational and pedagogical conditions responsible for their implementation, etc. conducting an expert assessment of the forecast; neural networks; EDM			
5	the stage of developing recommendations; synthesis of forecasts	creation of a regulatory framework in the field of forecasting on the basis of the university; informal methods			During the defense of this dissertation

Stages		Organizational and pedagogical conditions	Possible sources of forecast errors	Ways to troubleshoot errors	Application time
	V. Decision-making stage				
1	the stage of making decisions based on the results of forecasting; introduction of technology into the practice of construction universities	application of the whole complex of the above conditions; application of the whole complex of the above methods and means			after defending the thesis

Table 2.
Organizational and pedagogical conditions for predicting the educational results of students in construction areas of training.

- If pedagogical factors is developed on A-level AND psychological factors is developed on A-level AND social factors is developed on A-level THEN prognostic competence is developed on A-level

- If pedagogical factors is developed on F-level AND psychological factors is developed on B-level AND social factors is developed on A-level THEN prognostic competence is developed on D-level

4.6 Rules and methodology of modeling of dependence of prognostic competence general level depend on selected teaching model and frequency of application of prognostic activity

Every person at least once a day makes plans for the next day. Therefore, as a very low frequency of predictive activity, we will take 7 times. If on each pair 2–3 problems of predictive nature are solved, then about 50 prediction problems will be solved. This frequency is taken as a very high level of the frequency of predictive activity. Conventionally, we distinguish the following frequency levels (once a week): very low (FX), low (F), below average (E), medium (D), above average (C), high (B), very high (A). The method is trimf.

We will designate prediction learning models as follows: lack of special prediction learning – FX, non-systemic prediction learning – F, semiotic educational model – D, simulation educational model – B, social educational model – A. Method – trimf. The range is from 0 to 100 conventional units.

The creative level of the development of predictive competence will be estimated at 100 points (A), productive – 75 points (B), reproductive – 60 points (D), formal level – below 60 points, but above 20 (F). The type of membership function is gaussmf.

We have made 16 rules of fuzzy logic. Some of them are presented below:

- IF the prediction activity frequency is in A-level AND the training B-model is applied to THEN the predictive competence is developed on A-level;

- IF the frequency of prediction activity is in E-level AND the training B-model is applied to THEN predictive competence formed on D-level.

4.7 Forecasting technology

Forecasting is a stepwise process. The developed system of organizational and pedagogical conditions for each stage of forecasting is presented in the table format (**Table 2**).

5. Results

The result of modeling in the format of 3D projections of multidimensional dependence of general level of prognostic competence are presented in **Figure 2**. The projections onto the 2D plane are shown in **Figures 3–6**.

We see the complex nature of the dependence at all intervals. Competence is an integrative quality of a person, therefore it is more correct to interpret it according to the totality of all known factors, and not to look for the dependence of only some indicators on others.

The dependence of the formation of competence on the cognitive component has a stepwise nature, i.e. on a number of intervals, it practically does not depend

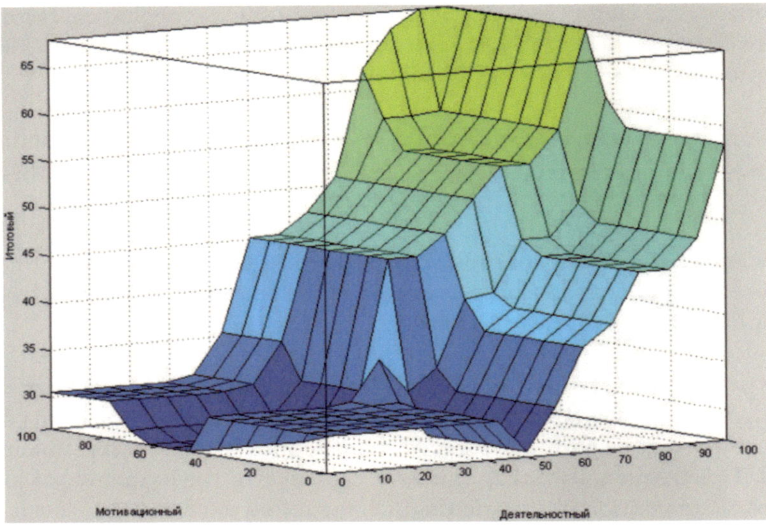

Figure 2.
The dependence of the final value of the level of formation of predictive competence on the level of formation of motivational and activity components.

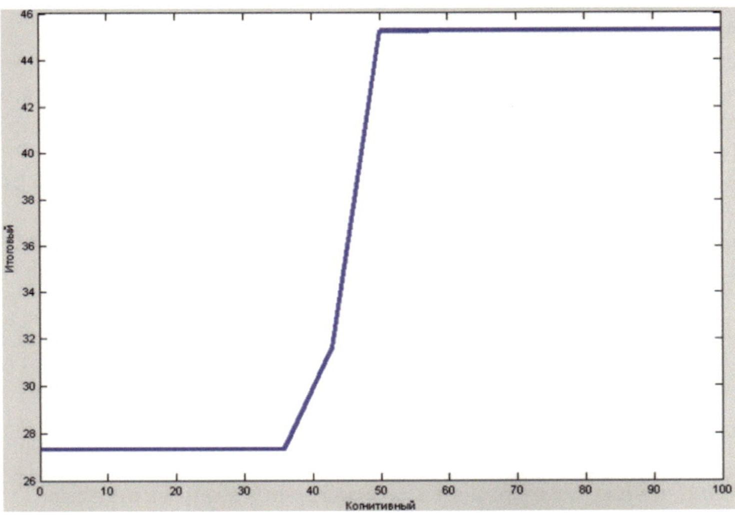

Figure 3.
The dependence of the final value of the level of development of prognostic competence on the level of formation of the cognitive component.

on external conditions and efforts of both students and teachers, but in the middle of the range, the level of competence is extremely dependent on cognitive efforts, quantitative characteristics dialectically turn into qualitative ones, so teachers should be as attentive as possible.

The dependence of the formation of competence on the activity component (**Figure 4**) is of the most complex nature, and advice can be given at each section of the level of formation. So, in the section from 30 to 50%, there is a maximum increase in qualitative changes with an increase in the number of solved prognostic problems, but then the slope angle decreases, although in any case the more we do prognostics, the greater the positive effect is observed. Every effort must be made to go from 90–100%.

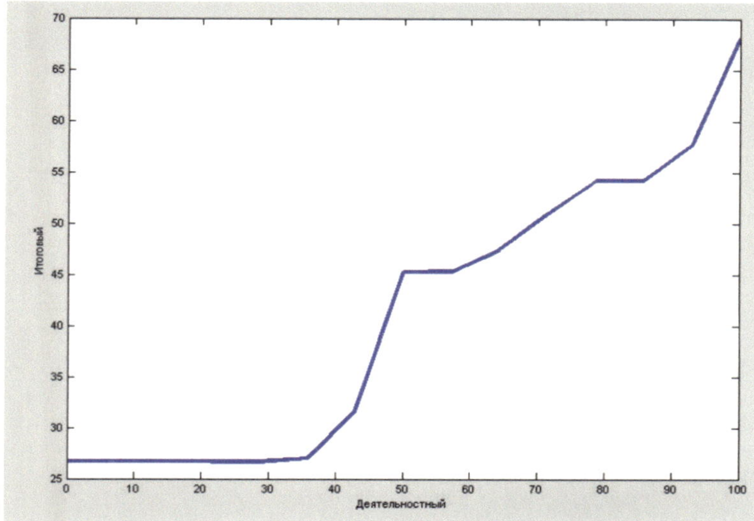

Figure 4.
The dependence of the final value of the level of formation of predictive competence on the level of formation of the activity component.

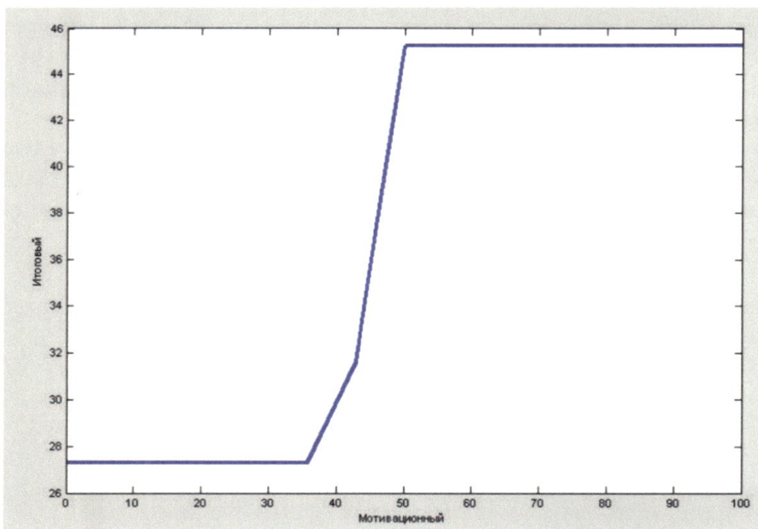

Figure 5.
The dependence of the final value of the level of formation of predictive competence on the level of formation of the motivational component.

Motivation for forecasting also has a "stepwise" effect: a low level of motivation is identical to its absence, it is impossible to distinguish between the level of motivation "above average" and "high". There is a dichotomy: the student is motivated - he learns to forecast, not motivated - he is not taught.

As in the previous graphs, we see a stepwise level of dependence of futurological skills on sociability. The student is able to form the skill of forecasting only in communication: with the teacher and classmates.

The result of modeling of development of prognostic competence depend on the most factors is presented on **Figures 7–11**.

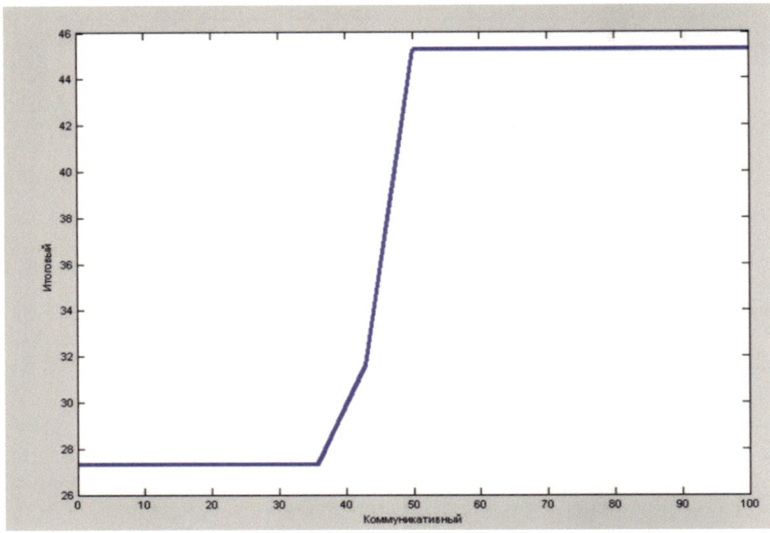

Figure 6.
The dependence of the final value of the level of formation of prognostic competence on the level of formation of the communicative component.

Pedagogical factors have the most important influence on the level of competence formation. Transitions from reproductive to productive, and from productive to creative are not possible without a mentor, although within these levels the teacher can have only minimal influence. Since there is a "plateau" on the graph.

Forecasting can be taught to almost any student, but for a qualitative transition to a new level of proficiency in futurology, you need to create a special psychological climate, so without a mentor it is difficult to independently learn predictions.

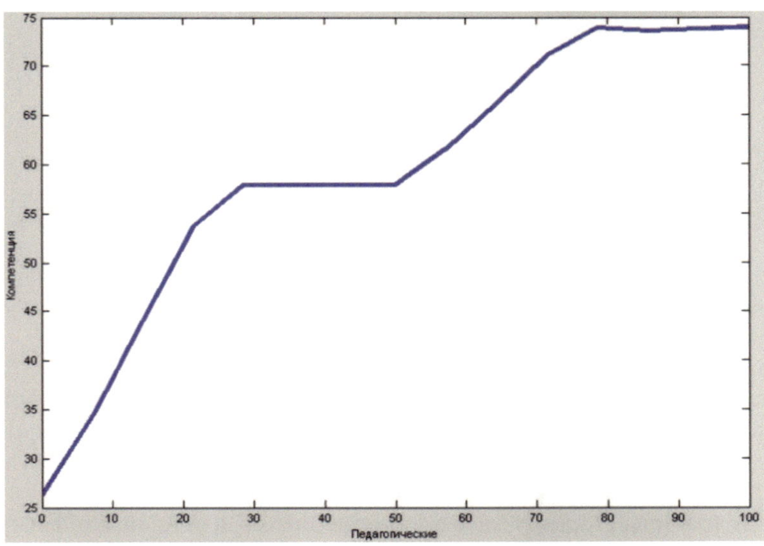

Figure 7.
The dependence of the formation of prognostic competence of pedagogical factors.

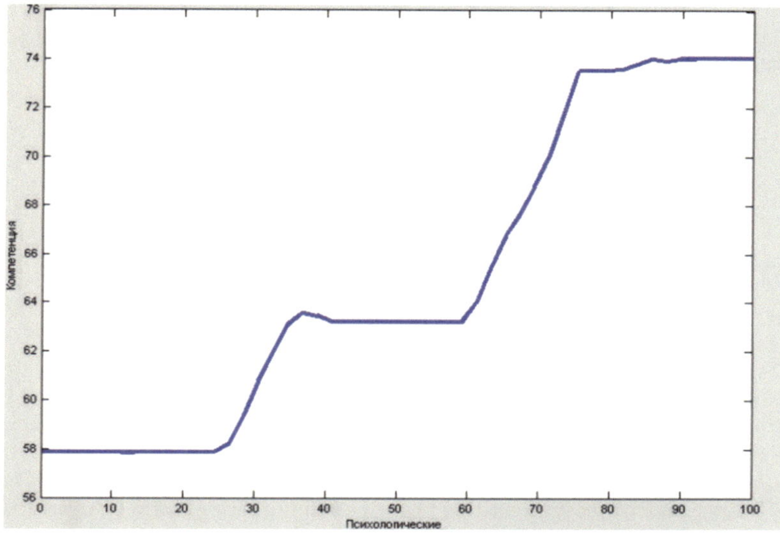

Figure 8.
The dependence of the formation of prognostic competence of psychological factors.

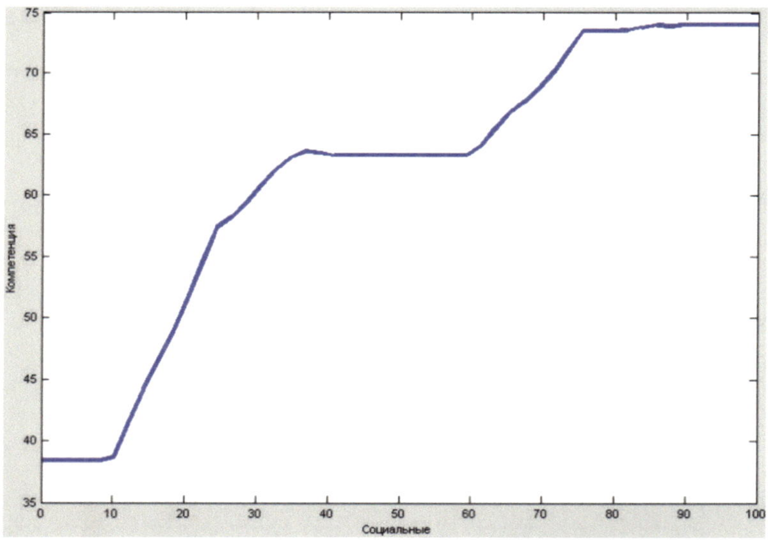

Figure 9.
The dependence of the formation of prognostic competence of socio-economic factors.

A low level of socio-economic security means a poor skill in the formation of predictive competence, and any minor improvements in the economic component significantly improve futurological skills. However, with an average level of social security, there is an insignificant plateau, at which an increase in the economic well-being has practically no effect.

As indicated in the previous modeling, in **Figure 4**, the activity has a significant improvement in predictive skill at all intervals. Thus, the two different models confirm each other.

Up to the achievement of a productive level of prognostic competence by a student, all teaching methods are equally effective, but in further training,

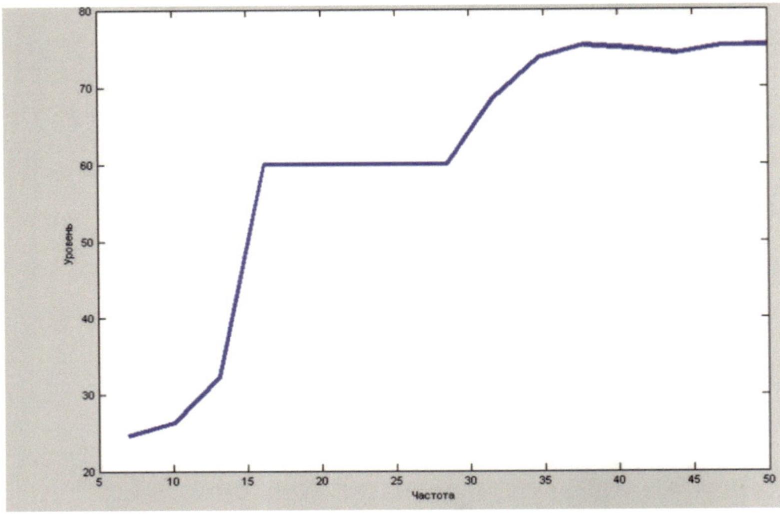

Figure 10.
The dependence of the formation of prognostic competence on the frequency of application of prognostic activity.

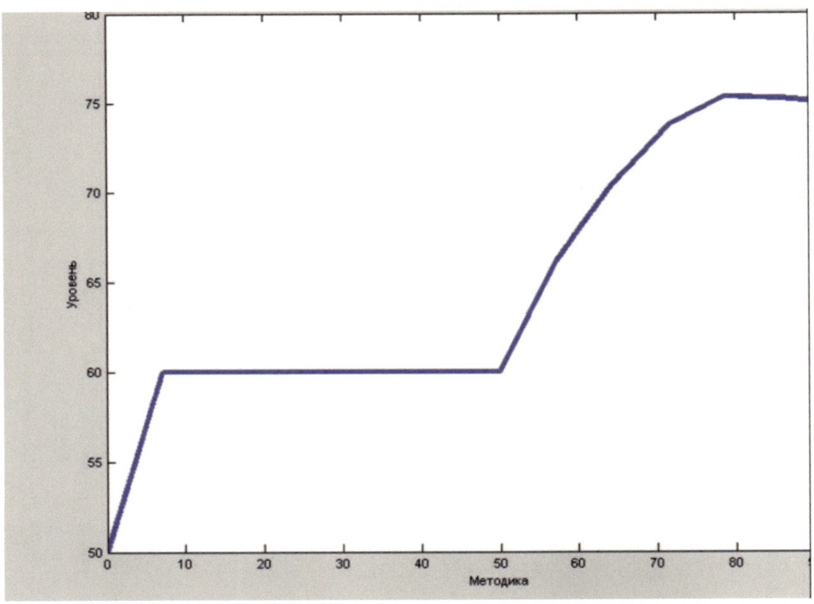

Figure 11.
Dependence of the formation of the prognostic competence of the technique used.

the selection of the correct individual trajectory for teaching foresight skills is extremely important, because any minor changes in the teacher's behavior model can lead to synergistic changes ("bifurcation point") in learning outcomes. One should "sensitively" feel all the changes in the student's achievements, because the dependence has a rather complex non-linear character.

6. Discussion

Most of the level of development of prognostic competence depends on the formation of its activity component. This is consistent with Selivanov's statement [12] the main means of education is the activity of the one being educated. The nature of dependence on each component is non-linear. We conditionally accepted that the formal level of development of predictive competence is less than 60%, but also more than 20%. This is due to the fact that for all students, the ability to predict is formed to one degree or another, but it cannot be stated that a student has a sufficient level of competence for engineering activity with a forecast accuracy of less than 60%. But each of the components formed at a sufficiently high level with the underdevelopment of other components of the competence is not a reason to assume that the prognostic competence is developed at least at the reproductive level. The same can be said about the situation when one of the components is not developed absolutely. In this case, the general level of development of predictive competence is not zero.

At a number of intervals in **Figures 7–10** various factors have virtually no effect on the formation of prognostic competence. We can influence only pedagogical factors, and the teacher cannot influence other factors directly, only indirectly. The greatest tangent of the angle of inclination of the dependence of the formation of prognostic competence on pedagogical factors is observed in the interval of the effectiveness of their use, which corresponds to the "above average" estimate.

The level of development of prognostic competence depends both on the frequency of application of prognostic activity and on the applied teaching methods, but nonlinearly. If you do not engage in prediction skills with students, prognostic competence will not be formed. With a very frequent prediction exercise, but without system lessons using pedagogical technologies, only a reproductive level of competence can be achieved. At a number of intervals, both the frequency of application of prognostic activities and the change in the effectiveness of the methodology have practically no effect on the formation of competence.

7. Conclusion

Prognostic competence is an integrative skill that is associated with many factors. Most of all, the level of development of prognostic competence depends on the formation of its activity component. Intellectual systems are a powerful tool for modeling the development of the prognostic competence of a civil engineer. The teacher can have the greatest influence on students who have achieved average achievements in the field of forecasting, but its influence on the most and least successful students is less effective.

We've discussed some of the problems have been reviewed at our last papers, for example [13–17] and others.

Author details

Juriy Andreevich Tashkinov
Donbass National Academy of Civil Engineering and Architecture,
Khartsyzsk, Ukraine

*Address all correspondence to: j.a.tashkinov@gmail.com

References

[1] Gerald A. Straka. Measurement and evaluation of competence. The foundations of evaluation and impact research. Descy, P.; Tessaring, M. (eds), 2004. – 51p.

[2] Bestuzhev-Lada IV Social forecasting/IV Bestuzhev-Lada, GA Namestnikova. Moscow: Pedagogical Society of Russia, 2002. 269 p.

[3] R.S. Baker. Educational Data Mining and Learning Analytics / R.S. Baker, P.S. Inventado // Learning Analytics. – 2014. – pp. 61-75. – doi: 10.1007/978-1-4614-3305-7_4.

[4] Dutt A. A Systematic Review on Educational Data Mining [Электронный ресурс] / A. Dutt, M.A. Ismail and T. Herawan. – vol. 5. – 2017. – pp. 15991-16005. – doi: 10.1109/ACCESS.2017.2654247.

[5] Romero C. Handbook of Educational Data Mining / C. Romero, S.Ventura, M. Pechenizkiy, R.S.J.D. Baker. – New York: CRC Press, 2011. – 526p.

[6] E.V. Calko. Predictive competence of future cultural workers/General and Professional Education. – 2013.- Vol.3, pp. 26-30.

[7] A.M. Abdullina, A.R. Kamaleeva, N.S. Galimov. The model of formation of professional competencies of students as future engineers in the process of learning the course "Theoretical Mechanics" at the university. – Kazan: Kazan Pedagogical Journal. – 2013.

[8] V.V. Bilyk. Experimental verification of the effectiveness of the formation of the design competence of the future engineer-teacher sewing profile // Scientific and methodical electronic journal "Concept". – 2014. – Vol. 10 (October). – pp. 171-175.

[9] M.G. Kolyada, T.I. Bugaeva. Pedagogical forecasting in computer intelligent systems: A manual – Moscow: Rusayns publishing house, 2015. – 380 p.

[10] M.G. Kolyada, T.I. Bugaeva. Pedagogical forecasting: theoretical and methodological aspect. Monograph. Donetsk: Knowlick, 2014. 268 p.

[11] L.I. Mamonova. Factors affecting the formation of general professional competencies of university students // Fundamental research. – 2012. – Vol. 6-2. – pp. 365-368.

[12] B.C. Selivanov. Fundamentals of general pedagogy: Theory and methods of education: Textbook. allowance for stud. higher ped. studies, institutions. – Moscow: "Academy", 2000. – 336 p.

[13] Ju. A. Tashkinov, M.G. Kolyada. Computer technologies in the educational process of construction university students (on the example of chemistry classes) / Interactive technologies in modern education: a collection of scientific papers. – Orsk: Publishing house of Orsk Humanitarian-Technological Institute (branch) of OSU, 2018. – 143 p. – ISBN 978-5-8424-0911-2 (pp.112-117)

[14] Ju. A. Tashkinov. Modeling the formation of a general level of prognostic competence of a civil engineer / Donetsk Readings 2018: education, science, innovation, culture and modern challenges. – Donetsk: DonNU Publishing House, 2018. – 418 p. (pp. 181-183)

[15] Tashkinov Ju.A. [Future Civil Engineers Educational Outcomes Predicting Synergetic Model]. Pedagogicheskoe obrazovanie v Rossii [Pedagogical Education in Russia]. 2020, No. 6, pp. 146-155. DOI: 10.26170/po20-06-17.1.

[16] Tashkinov Ju.A. [Forecasting Construction Engineering Students' Learning Outcomes by Means of Computational Pedagogys]. Integratsiya obrazovaniya [Integration of Education]. 2020; 24(3):483-500. DOI: https://doi.org/10.15507/1991-9468.100.024.202003.483-500.

[17] Tashkinov Ju.A. Pedagogical forecasting of educational results of future civil engineers in real time. Personality in a changing world: health, adaptation, development. 2020, Vol. 8 No. 1 (28). pp. 35-45. DOI: 10.23888 / humJ2020135-45.